CURIOSITY
GUIDES

THE HUMAN GENOME

CURIOSITY
GUIDES

THE HUMAN GENOME

The Book of Essential Knowledge

John Quackenbush, Ph.D.

Foreword by John Sulston, Ph.D., Nobel Laureate

imagine!
Publishing

An Imagine Book
Published by Charlesbridge
85 Main Street
Watertown, MA 02472
(617) 926-0329
www.charlesbridge.com

Library of Congress Cataloging-in-Publication Data
Quackenbush, John.
Curiosity guides : the human genome / John Quackenbush.
    p. cm.
Includes bibliographical references and index.
ISBN 978-1-936140-15-2 (hardcover : alk. paper)
1. Human genome--Popular works. 2. Genetics--Popular works. I. Title.
QH431.Q83 2010
611'.018166--dc22
                    2010001395

(hc) 10 9 8 7 6 5 4 3 2 1

Illustrations created in Adobe Illustrator by Haude Levesque
Display type and text type set in Weiss, Garamond, and Granjon
Manufactured in China, October, 2010
Designed by Linda Kosarin / theartdepartment.biz

To my son, Adam, who as a young child
represents the unlimited potential that lies deep within all of us
and goes beyond the genes encoded in our genomes.

JQ

CONTENTS

⬥

ACKNOWLEDGMENTS

I want to thank my wife, Mary Kalamaras, for letting me steal away the hours required to tell the tale of the human genome. As my spouse, Mary gave me the support and encouragement to write this book. As also my editor and most critical reader, she helped me shape my work into something that I hope readers will find both informative and entertaining.

Special thanks go to designer Linda Kosarin, whose talent, hard work, and wonderful creative energy are evident throughout the pages of this book, and whose grace and sense of humor were most appreciated throughout the design process. I am also grateful to Haude Levesque for the diligence and patience she displayed in creating (and re creating) the illustrations.

Finally, I would like to thank Charlie and Jeremy Nurnberg of Imagine Publishing for their vision of the *Curiosity Guides* and for their guidance in moving this volume from concept to publication.

JQ

FOREWORD

♒

In 1992, a small group of us founded a new organization, the Sanger Centre (now known as the Wellcome Trust Sanger Institute), dedicated to making a start on the sequencing of the human genome. As our institute grew, we, along with other scientists around the world, had the good fortune to work on one of the most exciting scientific projects of our time—the Human Genome Project.

In 2000, sitting in London at the Prime Minister's official residence, I heard Bill Clinton and Tony Blair inform the world that we now had a new map that would help to change the way we view ourselves and how we understand health and disease.

Although a milestone had been achieved, this was in truth a beginning, not an end. The sequence itself was as yet unfinished, but crucially it was in the public domain. This openness allowed finishing to proceed cooperatively, but above all it allowed everyone to access the genome freely for analysis and use.

Since then, this sequence assumes ever greater importance as it is analyzed in depth, as numerous other genomes are sequenced and compared with it, as new gene families are discovered, and as biologists use all this information to grapple with the enormous complexity of the processes of life. Genomics is embedded in and illuminates all of biology, and it is increasingly becoming a part of our everyday lives.

This is what makes *The Human Genome* by John Quackenbush a timely and important book for anyone interested in learning about the history, science, and social and ethical implications of the Human Genome Project. In these pages you will find a wealth of information, presented in a concise and accessible manner, that provides the necessary foundation for understanding genetics, genomics, and the technological advances that are changing not only the face of medicine, including the diagnosis and treatment of disease, but society as a whole.

Now that we humans have access to our genome sequence, it is imperative that we recognize the tremendous responsibility that goes along with such knowledge and the consequences that can arise from its application. *The Human Genome* will help you understand how and why genomics is changing the world and what the code of instructions written into every one of our genomes has to say about who we are as individuals and as a species.

JOHN SULSTON, PH.D.
2002 Nobel Laureate
Chair, Institute for Science, Ethics and Innovation
The University of Manchester
Manchester, U.K.
Former Director, Wellcome Trust Sanger Institute
Hinxton, U.K.

A Race to the Future

On June 26, 2000, before a gathering of U.S. and international dignitaries in the East Room of the White House, President Bill Clinton held a press conference to announce an astounding scientific milestone. After setting the stage by recalling Lewis and Clark's historic mapping of the American frontier, the President proclaimed the completion of another, very different type of exploratory map: the first survey of the entire genetic makeup, or *genome*, of the human being, consisting of all the DNA that our cells carry.

"Without a doubt," Clinton remarked, "this is the most important, most wondrous map ever produced by humankind." British Prime Minister Tony Blair, who had joined the President in the announcement via satellite link, further described the feat as "the first great technological triumph of the twenty-first century."

Clinton then spoke about the profound implications of the

achievement and the potential for medical advances in the detection, treatment, and possible cure one day of thousands of diseases, including Alzheimer's, Parkinson's, diabetes, and a number of cancers. Although genetic analysis had been ongoing for years and yielded many successes, including sophisticated drugs for certain types of leukemia and breast cancer, the mapping of the human genome promised to usher in a revolutionary new era in biology and medicine. Not surprisingly, the scientific community, along with the world at large, greeted Clinton and Blair's announcement with excitement and began imagining a future of seemingly limitless possibilities.

Standing alongside the President during the press conference were Francis Collins and J. Craig Venter, rival scientists who had each headed separate public and private efforts to sequence the genome. As director of the National Human Genome Research Institute (NHGRI) at the National Institutes of Health (NIH), Collins had led the publicly financed Human Genome Project (HGP), an international effort to discover all the genes in the human genome and determine the order of the billions of DNA bases they comprise. NHGRI was the agency charged with overseeing the sequencing and ensuring the effective use of the resulting information.

Venter was president of Celera Genomics, a private enterprise founded several years after the start of the HGP, in large part to independently sequence the human genome by 2001, four years earlier than the government's target. By declaring its ambitious goal and timeline, the company set off one of the most feverish races in the history of modern science. To better understand the circumstances and stakes involved, it is worth looking back at the events leading up to that celebratory summer day in Washington, D.C.

A Brief History of the Human Genome Project

It is difficult to pinpoint the beginning of the Human Genome Project. As with most scientific endeavors, the project was built upon decades of research and the contributions of hundreds, if not thousands, of scientists from around the world. Formally speaking, however, the project's generally agreed-upon start date is 1990, when the U.S. Department of Energy (DOE) and the NIH presented a joint plan to Congress outlining a bold vision for sequencing the human genome. It was the culmination of years of meetings, reports, and pilot projects, undertaken mostly in an atmosphere of healthy competition.

In the United States, the NIH has historically taken the lead in supporting research in human health and disease, commanding the lion's share of government funding for biology. However, when it came to its ongoing genetic research efforts, the NIH tended to fund projects that focused on studying one gene at a time, attempting to uncover each individual gene's makeup, function, and link to disease. So at least initially, the NIH was not interested in undertaking something as large-scale as the HGP.

In contrast, much of DOE funding went to physics research, giving the agency a long-established reputation for coordinating multidisciplinary projects that often involved thousands of scientists. But because of an interesting quirk in the DOE's history—

its roots in the former Atomic Energy Commission—it also had a record of genetic research into the effects of ionizing radiation on human health. At some point, DOE leaders recognized an opportunity to apply a physics-like "big-science" approach to producing a comprehensive roadmap of the human genome that would accelerate biomedical research.[1]

In March 1986, the DOE's Office of Health and Environmental Research (OHER) held a workshop in Santa Fe, New Mexico, to assess the feasibility of producing a reference sequence of the human genome.[2] The workshop findings confirmed that despite significant scientific, technical, and financial challenges, there was sound scientific justification to attempt the project and a reasonable expectation that the required technologies could be developed. Participants also agreed on including an educational and social component that examined the project's promises and limitations and the potential ramifications of making genomic information available.

Following the workshop, the DOE announced its Human Genome Initiative and provided seed funding for developing sequencing technologies. During the next few years, the agency ramped up its efforts and, through a series of research projects and meetings (including congressionally mandated advisory meetings), its human genome program began to take shape.

Not wanting to be left behind, the NIH partnered with the DOE, and through a series of scientific papers and presentations to Congress, the agencies successfully founded their $3 billion, multiphased research program for mapping and sequencing the genome by 2005. The NIH would eventually assume leadership of the project, creating the National Center for Human Genome Research (the precursor to the NHGRI) and putting Nobel laureate James Watson, the codiscoverer of the structure of DNA, at its helm. With a highly acclaimed and widely respected scientist leading the effort, the HGP had immediate scientific credibility.

Of course, the project was not solely a U.S. initiative. Scientists, government agencies, and private foundations around the world quickly signed on. Given the magnitude and complexity of the problem, however, international cooperation and coordination was essential. One of the leaders in bringing together all the diverse groups involved was the Wellcome Trust, a U.K. private charity that provided funding for the project at a level second only to that of the U.S. government.

As with other big-science projects, the HGP was initially guided more by a vision than a roadmap. But its goals were clear: identify all the genes in human DNA; determine the sequences of the three billion base pairs that make up the human genome; store the resulting data in public databases; improve data-analysis tools; transfer genomic technologies to the private sector; and address the ethical, legal, and social issues that might arise from the HGP.

During the 1990s, the HGP was marked by progress and inertia, excitement and frustration, as independent groups with independent approaches and opinions tried to work together toward a shared set of goals. It was into that atmosphere that Celera Genomics would make its debut as the ambitious private competitor, with Venter leading the way. Ironically, prior to assuming leadership of Celera, Venter had at one time worked at the NIH; after leaving, he had also received grant funding from the NHGRI to sequence parts of the human genome.

Celera's involvement was disconcerting to many scientists given its decided advantage of using public resources and data, to which it would add its private data. And although it was unclear to many how Celera would use the genome data, nearly everyone in the public HGP believed it would not be made freely available. This was counter to the NHGRI's mandate; which was to make the entire sequence available to everyone.

As the public and private interests each pressed on, both sides jostled for position and, in the process, made their own unique contributions to the scientific effort. In the end, Clinton's press conference would signal a public-private "tie" in the deeply con- tentious race. It was a bit of political diplomacy that allowed both groups to share in the achievement and to save face as they con- tinued their work, which in actual scientific terms was unfinished (and would remain so until 2006). Technicalities aside, what mat- tered most was that the first working draft of the human genome had been completed. It was now time to figure out how to use it to advance medical science in the ways promised when the HGP began and alluded to by President Clinton in his speech.

About the Map

Although Clinton spoke of "the" map of the human genome, there were actually two versions—one from Celera and one from the HGP. Each had different gaps as well as slightly differ- ent structures and collections of genes. Regardless, they were far more alike than dissimilar, and each was more than 95 percent complete.

The President's comparison of the technological feat to Lewis and Clark's survey of the American West was apt in that both resulting maps essentially sketched out a general "landscape" for the purposes of further exploration. The expeditioners' map served as an important rough guide and catalyst for western expansion. Likewise, the first working drafts of the human genome sequence facilitated the expansion of another kind of frontier, that of genomic research and, with it, the quest to uncover the nature of human disease, develop new diagnostic and prognostic medical tests, discover effective therapies, and establish a level of personalized medicine undreamed of even a decade ago.

Surprisingly, the driving force behind this revolution in biology and medicine has not been the genome sequence itself, but the technological advances that have emerged from the HGP. Scientists are using these tools to examine, at the molecular level, the roles that genes, the genome and its structure as a whole, and environmental factors play in human disease. More exciting are the technologies that will enable everyone to know most if not all of their genome sequence, not at a cost of billions of dollars as was true of the first genome, but for less than the price of an old used car. They also promise to radically change how medicine is practiced. We are fast approaching a time when doctors will be able to examine their patients' genetic profiles, determine their unique genetic characteristics and susceptibility to certain diseases, and design treatments to prevent or cure them.

Naturally, this raises the question of what the human genome sequence will mean for everyone in the long run. We all have a stake, and a responsibility, in how this brave new world unfolds because the information stored in our genome is in many ways the most personal information there is. As such, the implications for exploring the link between our genomes and our health go well beyond science and medicine and extend into public policy and civil rights. What must go hand in hand with research is vigilance over how genomic advancements affect the day-to-day lives of ordinary people in their workplaces, in their doctors' offices, and in their private lives. The best defense against misuse of this science will be to understand our genomes and the information they encode.

Caveats aside, perhaps the single most important revelation that has thus far come out of studying the human genome is that, although the reference sequence does not reflect all the inherited gene variants that we possess individually, we are more similar to one another, and more intricately tied to the web of life on earth, than anyone previously imagined.

INTRODUCTION NOTES

1. In the United States, when the end of the Cold War was in sight, DOE officials recognized that Congress would cut funding for physics research, which had been justified up until then as being necessary for a strong national defense. Many believe that to avoid budgetary reductions, the DOE sought out a project too ambitious to be eliminated.

2. Since differences between the genomes of any two people were known to be extremely small, the completed genome sequence would serve as a resource for every human being.

CHAPTER 1

⟨ornament⟩

Back to Basics

To fully comprehend why the sequencing of the human genome was viewed as such an important advance—hailed by presidents and prime ministers—and how it is changing biological and medical research, it helps to explore the history of genetics, the essential concepts of molecular biology, and the discoveries that led up to the Human Genome Project.

When trying to understand how a body of scientific knowledge comes to be, some people might liken the process to the construction of a tower of blocks, with each block representing a discovery laid atop the one before. But the better analogy would be the assembly of a jigsaw puzzle, with discoveries "fitting" together at different times and the entire scientific picture often not appearing until many pieces are in place. Likewise, the story of genetics, and of genomics, focuses on critical events and discoveries that often came together not chronologically, but in hindsight, to arrive at a greater scientific understanding.

MENDEL'S GARDEN: SETTING THE STAGE

You probably remember Gregor Mendel, the "father of modern genetics," from your high school biology class. An Austrian monk and biologist with a passion for—perhaps even an obsession with—breeding plants, Mendel noticed that in crossing different plants, the traits of parent plants were reflected in the offspring. Mendel's pea plants represented the first genetic model organism, a nonhuman species that helped him develop his *Laws of Inheritance*, principles that also apply to human biology. We know that traits in humans are passed from parents to children, and that nearly every newborn is scrutinized for their father's eyes or mother's mouth or Uncle George's unusual ears. But studying inheritance patterns in humans is difficult; a single generation spans about 20 years and the number of offspring we have is relatively small. The beauty of studying plants is that one can breed a new generation every year or more often, and offspring from a single plant can number in the tens or hundreds.

Mendel observed that not only were traits passed between pea plant generations, but there existed different types of traits depending on the genes inherited. For example, round peas dominated over wrinkly peas. And if Mendel crossbred plants that produced yellow peas with those that produced green peas, the offspring were likely to produce yellow peas. More important is that by examining hundreds of offspring, he discovered that inheritance patterns could be described using precise mathematical relationships, since offspring carried two possible variants for any given trait, having received one variant copy from each parent plant. These simple yet fundamental observations set the stage for modern genetics and the exploration of the role of inheritance in human disease.

DARWIN'S NATURAL SELECTION

Mendel's Laws of Inheritance, although elegant, raise an important question: If we are the product of the traits we inherit, then how has such a great diversity arisen among humans and other species? To a large extent, British naturalist Charles Darwin answered this question with his theory of evolution based on what he called *Natural Selection*. Darwin believed that in a world populated by a diversity of individuals, some would have physical characteristics making them slightly more "fit" than others to survive and, what is most important, more likely to reach reproductive age, attract a mate, and pass on their traits to offspring. Additionally, a species could be "remade" over time to accommodate environmental adaptation. The process he described was often a slow one occurring across successive generations, but it was not hard to imagine.

Consider the flying squirrel—a peculiar creature with wing-like flaps of skin stretching between its front and back legs. What led these animals to evolve such a strange physical attribute? Given the millions of years, or the thousands of generations, involved in evolution, the answer is fairly straightforward. Squirrels live in trees and falling from the top of a tree is hazardous to a squirrel's health. Squirrels with extra skin between their limbs generate more air resistance during their leaps, gaining extra "hang time" as they jump from branch to branch. If they live in an environment where this trait is advantageous (a place with tall, well-separated trees, for example), squirrels that glide are more likely than their non-gliding neighbors to survive and produce offspring. In fact, it is easy to see how squirrels possessing such extra skin could come to dominate the local squirrel population.

Why then are not all squirrels flying squirrels? It turns out that natural selection is a perfect example of how genetics is only one player in the evolution game. Environment plays a huge role in determining what drives natural selection. When environments change, they redefine what makes one individual more or less fit than another for survival. Gliding might be a superior trait for navigating one environment, but a squirrel living in another area populated by different trees or predators might benefit more from an altogether different trait. In other words, there is no such thing as the perfect squirrel.

Darwin's great revelation with regard to evolution occurred in 1835 when he visited the Galápagos Islands, an archipelago set in the Pacific Ocean and off the coast of Ecuador. There, he made numerous observations of island fauna, taking back with him many specimens of mammals and birds. But of these, it would be the various species of Galápagos finches that would become what we now consider as being the iconic representation of natural selection at work.

The Galápagos are a series of volcanic islands that arose devoid of life. Over time, they became populated by what washed ashore, what was carried to them by the wind, or what early human visitors brought with them, in the process evolving unique ecosystems comprising slightly different collections of plants, animals, and insects. And so, on each island, natural selection tilted the balance in subtly different ways to favor the finches best able to survive and reproduce within an island's particular ecosystem. On some islands, this led to finches having short, strong beaks that could crack open plentiful seeds. On others, the evolutionary winners were finches with long beaks, efficient for retrieving insects burrowed in plants.

Darwin observed that the process of natural selection was not limited to finches. Each island had different competing animal species and opportunities that, if exploited, could give certain ones an advantage in survival and, more important, in reproduction. Hence, the Galápagos Islands are an excellent example of how isolated ecosystems tend to evolve in their own way, with unusual creatures sometimes filling essential ecological niches.

But evolution is not necessarily confined to isolated environments. Bacteria can evolve to survive antibiotics, new strains of influenza appear each year, and plants, animals, and even humans can change along with their environments. A classic example of environmental influence involves England's peppered moth, which evolved from being light gray in color to dark gray as pollution from the Industrial Revolution increasingly darkened the landscape, making lighter-colored moths easier prey for birds.

By the end of the nineteenth century, thanks to passionate scientists like Darwin and Mendel, a strong foundation for the study of genetics had been laid. Mendel had elucidated the mathematics of how traits were passed among generations, and

Darwin had explained how certain traits could become common in populations and how environments could influence the process. But questions remained as to where traits resided in the cell and why diversity existed at all. In time, DNA would provide part of the answer. We will explore this shortly, but first, it is important to learn a few things about the structural and functional units that make up most living things.

CELLS: A HIDDEN UNIVERSE

We tend to think of scientific discovery as being driven by careful experiments, detailed observations, and the keen insight of scientists. But something equally important is involved: technology. The Human Genome Project was driven by the development of DNA sequencing machines, robotic systems for processing biological samples, and large-scale computer software for stitching together the sequence of the human genome. Technological advance has always paved the way for discovery throughout the history of biology, and one of the most well-known examples is undoubtedly the invention of the microscope.

Although the magnification properties of glass had been known since ancient times, it was not until around 1590 that a father-son pair of Dutch eyeglass makers, Hans and Zacharias Janssen, invented the forerunner to the compound microscope and telescope. While Galileo further developed the telescope in the early 1600s, using it to reveal planetary wonders, others improved the magnification and resolving power of the microscope to reveal the wonders of another, hidden, universe.

In 1665, physicist Robert Hooke published *Micrographia*, in which he chronicled his use of a microscope. Perhaps his most important observation was of cork that he described as an array of honeycomb-like compartments he called *cells* (from the Latin *cella*, for "small room"). It is not clear if Hooke realized the significance of this observation or its applicability to living things (cork is dead plant material), but the word "cell" would come to denote the basic unit of life.

Antonie van Leeuwenhoek, a Dutch shopkeeper considered to be the "father of microscopy," used his finely crafted microscopes to study yeast, blood, insects, and other small objects. In 1674, while examining pond water, he found it teeming with life. The "animalcules" that he observed, now referred to as *microorganisms*, were a diverse collection of tiny creatures, some single-celled (or *unicellular*) and others comprised of many cells (or *multicellular*). In Leeuwenhoek's time, the discovery of a world populated by creatures smaller than the eye could see came as a shocking revelation.

As microscopes and techniques for preparing cells for observation were refined, scientists learned that nearly all living things were composed of cells and that many cells contained other important structures.[1]

In 1831, Scottish botanist Robert Brown noticed that the opaque spots he observed in plant cells, now known as *nuclei* (the plural of *nucleus*), were common to many cell types. Soon after, scientists discovered many more subunits within cells. These *organelles*, which included *mitochondria*, *vacuoles*, and *chloroplasts*, had membranes separating them from the rest of the cell (think of them as discrete cellular "organs"), while other cellular components, such as the *ribosomes*, did not.

Different forms of life also had different types of organelles by which their cells could be classified. The simplest organisms are called *prokaryotes*. They are single-celled, lack a nucleus to house their DNA, and possess a single membrane (their outer cell wall). Prokaryotes include all bacteria and the recently discovered *archaea*.[2] Other single-celled organisms and multicellular organisms, known as *eukaryotes*, possess nuclei and other organelles and can be further classified based on their cellular structures. For example, only plants contain chloroplasts, the organelles that process sunlight energy into sugars.

These and other revelations arising from centuries of research led to the development of *cell theory*, a key principle of biology set forth in 1839 by physiologist Theodor Schwann and botanist Matthias Schleiden, and later revised by pathologist Rudolf Virchow. Among other tenets, cell theory states that cells are the fundamental structural and functional units of life, that all known living things are made up of one or more cells, and that new cells are formed from other existing cells.

Taking it further, a cell can be thought of as a little "machine" (made up mostly of proteins) that carries all the information necessary to make the specific proteins it requires to function. This information is passed from one generation to the next when the cell reproduces by dividing to create two daughter cells that, for all practical purposes, are identical to it. But how is this replication possible, and where is the information to accomplish it stored?

CHROMOSOMES AND GENETIC TRAITS

In 1875, improvements in microscopy allowed German biologist Walther Flemming to discover that the nucleus itself contained still smaller structures. Because they could be stained with dyes, Flemming named them *chromatin* (from the Greek *chroma*, meaning "color"). He noticed that when eukaryotic cells divided, the chromatin organized itself into stringy bodies that were later called *chromosomes* (see Figure 1).[3]

In 1902, American biologist Walter Sutton observed that during cell division, chromosomes grouped into similar-looking pairs that duplicated themselves before pulling apart as the cell divided into two daughter cells, each of which ended up with a full set of chromosomes. He also noticed that when sperm and egg cells formed, these sex cells, or *gametes*, carried only half the full set of chromosomes. It was not until a sperm fertilized an egg that the full set of chromosome pairs was re-created. This led Sutton to speculate that chromosomes somehow carried genetic traits.

It was geneticist and embryologist Thomas Hunt Morgan who, through his work with fruit flies, would finally validate Sutton's hypothesis. By then, scientists had known that in a specific chromosome pair in male humans and many other male animals, including fruit flies, one chromosome looked smaller and different from its partner. In females, the chromosomes of this same pair looked nearly identical. The unusual chromosome was dubbed "Y," while its partner was called "X." The discovery that females carry two X chromosomes and males carry one X and one Y would greatly aid Morgan in his study of how traits are inherited.

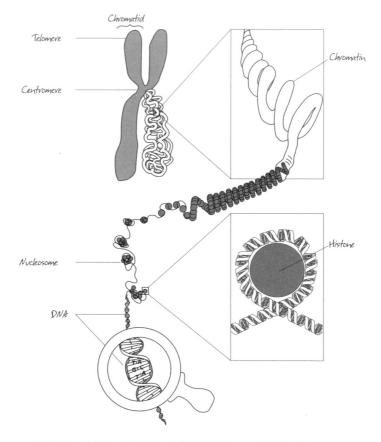

FIGURE 1: A HUMAN CHROMOSOME AND ITS STRUCTURE

A chromosome contains a single molecule of DNA, tightly coiled around proteins called histones and packaged into a compact, stringlike fiber called chromatin, to make up a structure called the chromatid. The centromere is what divides the chromosome into two sections, or arms, while telomeres are sections of highly repeated DNA sequences that protect both ends of the chromosome. Seen here is a representation of a chromosome during mitosis, when it is at its most highly condensed form.

The fruit fly (of the genus *Drosophila*) is a superb model organism to study, given its short life span and ability to create hundreds of offspring every generation. Although most fruit flies have red eyes, Morgan and his colleagues came across a male fly with white eyes and decided to use it to investigate how a trait could be passed from one generation to the next. When Morgan bred the white-eyed male with a red-eyed female, all the offspring were red-eyed, suggesting that, in fruit flies, red eyes are dominant over white eyes. When Morgan then crossed the flies of this first generation with one another, white eyes did appear, but only in the males. After careful data analysis, Morgan concluded that white eyes resided on the X chromosome. His discovery, corroborated by numerous other experiments in flies and other species, confirmed that genetic traits were carried exclusively on chromosomes. Morgan had uncovered the mechanistic basis of heredity.

HUMAN CHROMOSOMES

Every animal species carries a unique set of chromosomes, the numbers of which can vary. A laboratory mouse has 40 chromosomes, an elephant has 56, while a kingfisher bird, surprisingly, possesses 132 chromosomes.[4] Humans fall somewhere in the middle, with 46 chromosomes in 23 pairs. Twenty-two of these pairs are called *autosomes* and generally look alike under a microscope. Human chromosome pairs are numbered, based on overall size, from 1 to 22, with chromosome 1 being the largest and chromosome 21 being the smallest.[5] The remaining two chromosomes are the sex-determining chromosomes X and Y.

Today, we understand that changes in chromosomes and the number of them carried in each cell can lead to genetic disorders. For example, when human cells carry three copies of chromosome 21, a condition known as trisomy 21, or Down syndrome, occurs. Down syndrome is often associated with mild to moderate impairment of cognitive ability and physical growth, as well as a set of common facial characteristics. Other genetic abnormalities manifest themselves based on cells having too many or two few chromosomes, or only parts of chromosomes. They include trisomy 18 (Edwards syndrome), trisomy 13 (Patau syndrome), and monosomy X (Turner syndrome, where only one X and no other X or Y is present). These chromosomal disorders are rare, however, since fetuses that carry too many or too few chromosome copies generally do not survive to birth.

What these syndromes support in scientific terms is the idea that traits are carried on chromosomes and that the development of viable, so-called normal individuals requires a precise balance in the number of chromosomes in their cells. Yet because chromosomes are complex structures composed of nucleic acids and proteins, a question arises over what specifically within chromosomes carries traits.

DNA: The Molecule of Heredity

In the 1860s, about the time that Mendel was experimenting with plants, Swiss chemist Johannes Friedrich Miescher discovered that when cell nuclei were treated with a weak alkaline solution that was then neutralized with acids, a strange milky-white, stringy substance could be extracted that differed from the proteins known to exist in cells. This substance, which he called *nuclein*, was what we now know as *deoxyribonucleic acid*, or *DNA*.

DNA is not a simple molecule with a precise structure and fixed molecular weight like other molecules we learned about in high school chemistry. It is a polymer, a long molecule consisting of a string, or sequence, of nearly identical subunits organized like beads on a necklace. These subunits, interchangeably referred to as *nucleotides, nucleic acids,* or *bases,* are adenine, cytosine, guanine, and thymine, represented by the letters A, C, G, and T, respectively. Although each base differs slightly in structure, all four share a deoxyribose sugar "backbone" that binds them together in specific pairings, creating chains hundreds of millions of bases long. A specific DNA molecule is therefore often represented by a string, or sequence, of As, Cs, Gs, and Ts.

Although scientists in Miescher's time knew that cells seemed to carry a lot of DNA and that chromosomes appeared to be made of both DNA and proteins, they doubted that DNA was involved in heredity. An immunologist's experiments with bacteria would later change this view.

During the 1930s and 1940s, at New York's Rockefeller University, Oswald Avery was studying a process called bacterial transformation, whereby bacteria take up DNA from the environment through their cell walls. Avery had been building on previous experiments involving the bacterium *Streptococcus pneumoniae* that demonstrated how a nonvirulent form called "rough" could be transformed into a virulent form called "smooth." When he isolated and purified different compounds (proteins, lipids,

DNA) made from dead versions of the smooth type and grew them together with live versions of the rough, he discovered that only DNA transformed the bacteria. Despite initial doubts by scientists, Avery's results were later confirmed and the importance of DNA was acknowledged.

Scientists then wondered how such a simple molecule as DNA could carry complex genetic information. A number of groups tried to figure this out, but it was American James Watson and Englishman Francis Crick working at Cambridge University in the United Kingdom who ultimately prevailed on April 25, 1953, when they published a scientific paper in the British journal *Nature*, in which they described the structure of DNA.[6] Five weeks later, they followed up with another paper that explained how the structure allowed self-replication—essential if DNA were to be considered the molecule of heredity.

Watson and Crick's deduction was remarkably elegant. DNA was in the form of a double helix consisting of two nucleic acid strands running in opposite directions, with complementary bases paired together.

The double helix can be imagined as a spiral staircase with deoxyribose sugars forming the rails and pairs of bases joining together in the center to form the steps. In this model, which draws on countless other scientific observations about DNA, the base cytosine on one strand always pairs with guanine on the other (C with G). Similarly, adenine always pairs with thymine (A with T).

Remarkably, the helical structure also explains how DNA replicates itself to carry information during cell division (see Figure 2). As

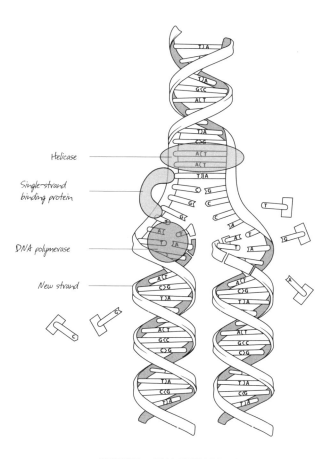

Helicase

Single-strand
binding protein

DNA polymerase

New strand

FIGURE 2: DNA REPLICATION

The process of DNA replication is quite complex, involving numerous enzymes. Helicases unwind the parent helix, single-strand binding proteins stabilize the separated strands whose hydrogen bonds have been broken, and DNA polymerase binds to the single template strands and synthesizes a complementary DNA strand. As copying is completed, hydrogen bonds re-form between the new base pairs, resulting in two identical copies of the original DNA molecule.

a cell prepares to divide, the DNA double helix unzips, separating into two strands of nucleotides. Following base-pairing rules, an enzyme called *DNA polymerase* reads the individual strands, using them as templates to both capture nucleotides made by the cell and to synthesize two new DNA strands that are identical to the original partner strands and complementary to the template strands. In this way, the DNA copies itself so that each daughter cell receives one complete DNA copy. In one fell swoop, Watson and Crick had the answer for a question dating back to Mendel and his peas: DNA carries the blueprint of heredity that is passed from one generation to the next.

The Central Dogma of Molecular Biology

Although scientists were incredibly excited over Watson and Crick's discovery, they still had questions about the nature and function of DNA, including how it could encode our traits—from eye color and blood type to the chance of developing cancer or even one's propensity to take risks.

A seemingly complex yet elegant answer would eventually emerge: the *Central Dogma of Molecular Biology*. This principle is often summarized as "DNA to RNA to protein" and refers to the flow of genetic information that occurs in a biological system. To understand how important this principle is to biology and to the HGP, it helps to review it.

We know that the proteins a cell manufactures ultimately define cellular traits, whether those of a neuron, a skin cell, a white blood cell, or other cell type. To serve as the molecule of heredity, DNA must carry a code that lets a cell make proteins. The problem is, compared to the final product—the proteins themselves—DNA seems to have less of a capacity to carry information, given its composition of only four bases. Proteins, however, are *polypeptides*, long-chain polymers of variable lengths that are composed of a combination of 20 different peptide, or *amino acid*, subunits.

Common sense dictates that more combinations can be derived from working with 20 units of something versus only four units. For example, if you tried to write a book using an alphabet of only four letters, it would be extraordinarily difficult to create all the words you would need. An alphabet of 20 letters would make the task significantly easier.

So how are proteins, composed of 20 amino acids, encoded using instructions that are "written" using only four DNA bases? Rather than rely on single DNA bases ("letters") A, C, G, and T, cells use three-letter base combinations called *codons* as instructions for building a protein.

Why three-letter combinations? Using single DNA bases allows only four possible choices, too few for protein manufacture. Two-letter combinations yield four options for the first base and four for the second base, a total of 16 base combinations— still too few. But three-letter combinations offer four times four times four letters, or 64 possible letter combinations, representing more than enough codons to string together to encode 20 amino acids (see Figure 3).

You might wonder if the many "extra" codons that can be created out of the DNA alphabet are ever used. This embarrassment of riches is actually put to good use. First, there is some

Alanine (Ala/A)	GCU, GCC, GCA, GCG	**Leucine** (Leu/L)	UUA, UUG, CUU, CUC, CUA, CUG
Arginine (Arg/R)	CGU, CGC, CGA, CGG, AGA, AGG	**Lysine** (Lys/K)	AAA, AAG
Asparagine (Asn/N)	AAU, AAC	**Methionine** (Met/M)	AUG
Aspartic acid (Asp/Dv)	GAU, GAC	**Phenylalanine** (Phe/F)	UUU, UUC
Cysteine (Cys/C)	UGU, UGC	**Proline** (Pro/P)	CCU, CCC, CCA, CCG
Glutamine (Gln/Q)	CAA, CAG	**Serine** (Ser/S)	UCU, UCC, UCA, UCG, AGU, AGC
Glutamic Acid (Glu/E)	GAA, GAG	**Threonine** (Thr/T)	ACU, ACC, ACA, ACG
Glycine (Gly/G)	GGU, GGC, GGA, GGG	**Tryptophan** (Trp/W)	UGG
Histidine (His/H)	CAU, CAC	**Tyrosine** (Tyr/Y)	UAU, UAC
Isoleucine (Ile/I)	AUU, AUC, AUA	**Valine** (Val/V)	GUU, GUC, GUA, GUG
START	AUG	**STOP**	UAG, UGA, UAA

FIGURE 3: THE UNIVERSAL GENETIC CODE

There are 64 triplets of bases, or codons, in the genetic code, each of which encodes for one of 20 amino acids. The codon AUG both signals the start of translation and encodes for the amino acid methionine. Shown here is an RNA representation of the code.

redundancy built into the genetic code. Multiple codons can be used for the same amino acid as a kind of genetic insurance that allows some small changes in the DNA sequence to occur without affecting the final protein made. Second, extra codons let the cell create additional instructions, including a "start" command (represented by a codon that also doubles as the code for the amino acid methionine) and three "stop" codons.

DNA TO RNA TO PROTEIN

The conversion of DNA code into proteins involves processes called *transcription* and *translation* that reflect discrete cell functions: information storage and protein construction. In eukaryotic cells, DNA is densely packaged in chromosomes and stored in the nucleus, while protein manufacture happens at the ribosomes located in the *cytoplasm*, which is the bulk of the cellular material found outside the nucleus. In contrast, the DNA and ribosomes in prokaryotic cells coexist in the cell's cytoplasm, since prokaryotes do not have nuclei.

Transcription occurs while the DNA molecule is unzipped. Regions of DNA that contain blueprints for proteins, most often referred to as *genes*, are copied, or transcribed, into a polymer related to DNA called *ribonucleic acid*, or *RNA*.[7] This is done through *RNA polymerase*, an enzyme that binds to unzipped DNA and slides along it, making an RNA polymer that is complementary to the DNA. Unlike its cousin DNA, RNA is naturally single-stranded. It has a similar sugar-phosphate backbone, except that the sugar is ribose instead of deoxyribose.

RNA also has four bases like DNA, although the thymine

(T) occurring in DNA is replaced in RNA by uracil (U). Following transcription, any portions of RNA that do not code for proteins are removed, and the RNA is further altered to improve its stability. The resulting molecule is known as *messenger RNA (mRNA)*.

The mRNA is transported out of the nucleus to the ribosomes, which "translate" the code into proteins by binding to the mRNA, reading its sequence one codon at a time and stringing together the corresponding amino acids. The resulting proteins, which began in the cell as long, single-peptide chains, subsequently fold up into three-dimensional structures that make up the machinery of the cell and dictate its function. Some proteins build the structure of the cell. Others are the enzymes that carry out the biological reactions that keep the cell alive, such as breaking down sugars to provide the cell with energy or managing cell division. Still others function to regulate (start or stop the transcription and translation of) other genes.

This process of using the information encoded in the DNA to create the machinery of the cell is really at the heart of this chapter and summarizes much of what we know about the genetic code in a way that one could explain in a five-minute elevator ride. It also illustrates why the HGP was so revolutionary. By creating a catalog of (nearly) all human genes, it allowed us to look at the entire set of building blocks needed to make a human cell—and to better explore how the process can go awry.

For example, we have all heard the term *gene mutation*, which describes a change in the DNA code that can result from, among many things, the swapping of a single base with another or the addition or deletion of one or more bases during DNA replication, or from environmental assaults and other causes. Many changes are benign and do not affect the protein encoded. But some mutations do change the code so that a different protein— one too short, too long, or altered in its amino acids—is made. This can significantly affect the three-dimensional structure of the protein. Altering the shape of the protein can change its function, and a change in function can ultimately lead to disease.

It is important to note that the translation process by which a DNA message is turned into protein is nearly universal. Organisms from bacteria to humans and nearly everything else on the planet use the same code. This commonality suggests that nature tends to be conservative and does not reinvent something that already works. It also means we can study human diseases using organisms that, at first pass, seem rather unlikely subjects. We can even take advantage of the universality of protein synthesis to create protein-based drugs by inserting a human gene—say the gene for insulin or human growth hormone—into a bacterium and letting the bacterium's cellular machinery do the manufacturing work.

GENE MAPPING AND SEQUENCING BEFORE THE HGP

As the scientific community explored the genetics landscape throughout the twentieth century, it looked beyond studying model organisms like plants and fruit flies to focus increasingly on human diseases, and it did so by relying on a simple concept:

Just as traits such as red hair or blue eyes run in families, so do certain diseases.

To that end, scientists developed more and more tools to map traits, including diseases, to specific regions of DNA within the human genome. The most useful of these regions were the small segments of DNA that varied naturally among humans, but which were heritable in families. A number of techniques were created to identify the presence of these variant sequences and to determine their order along the chromosomes, which in turn made them valuable because their unique positions could serve as biological landmarks within the genome.

By tracking these genetic markers and their variants in people and determining how they were inherited, scientists could narrow down where in the human genome a particular gene associated with a disease might be located.

A good analogy for this strategy would be to picture yourself sitting in a television studio in front of camera monitors set on Golden Gate Park in San Francisco, Lincoln Park in Chicago, and Central Park in New York City. You have been asked to report on how often you see actor and director Woody Allen walking through one of the parks. Even if he is unlikely to walk in these parks very often, you are much more likely to see him in Central Park than in any of the others. And if you saw him there enough times, you might conclude that he probably lives in the neighborhood—although you would almost certainly not know his street address. If you wanted to map his location more precisely, you could set up additional cameras at various New York City landmarks to narrow down the neighborhood in which he resided. If you repeated the process enough times, with increasingly focused observations, you might eventually figure out his home address.

Over time, geneticists were able to track down the "neighborhoods" of many genes associated with diseases and, in some cases, even create diagnostic tests that could reveal a person's risk of carrying them or passing them on. But while the ability to map the general location of genes partially answered the "where" of human disease, it did not answer the "how" or "why." This required finding the genes themselves, determining their DNA sequence, and seeing whether that sequence found in people with a particular disease differed from the sequence found in people without the disease. Scientists struggled for many years to find a good way to do this and, although they had limited success, their methods were painstakingly slow and often imprecise, even for DNA segments only 100 bases long (much smaller than a typical gene).

In 1977, biochemist Frederick Sanger of the Medical Research Council in the United Kingdom arrived at a revolutionary method for determining the sequence of As, Cs, Gs, and Ts within a particular segment of DNA. The process, now known simply as *DNA sequencing*, transformed the way scientists could investigate human disease and its causes and greatly accelerated biological research and discovery. If a good candidate gene was found using disease-mapping techniques, Sanger's method could then be used to accurately determine its DNA sequence; and by comparing the sequence of a gene present in people with a disease to those without it, causative mutations could be identified (see Figure 4).

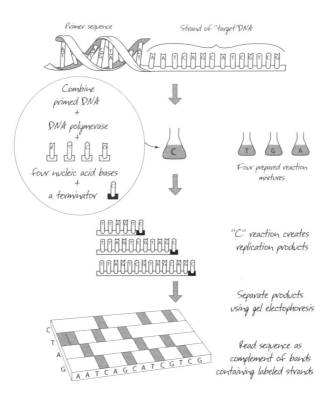

Primer sequence   Strand of "target"DNA

Combine
primed DNA
+
DNA polymerase
+

four nucleic acid bases
+
a terminator

C

Four prepared reaction mixtures
T G A

"C" reaction creates replication products

Separate products using gel electophoresis

C
T
A
G

A A T C A G C A T C G T C G

Read sequence as complement of bands containing labeled strands

FIGURE 4: THE SANGER SEQUENCING METHOD

In the Sanger method, a single "target" DNA strand is selected as a template for analysis and "primed" by a short stretch of complementary, double-stranded DNA lying "upstream" of the target region. The four nucleic acid bases being determined (C, T, G, A) require four separately prepared reactions. Each reaction contains: a DNA polymerase (to produce a DNA strand complementary to the target sequence), a mixture of all four bases, and millions of copies of the primed DNA target sequence. Each reaction also receives a terminator—a modified DNA base that stops the addition of bases to a growing DNA strand. In the C reaction, for example, millions of DNA strands are synthesized, but all are partial and eventually stop at "C." The T, G, and A reactions are similar with their own respective terminators. After the reactions are run, the synthesized DNA is removed and each reaction is analyzed using gel electrophoresis, a process that separates DNA molecules based on size. The DNA segments produced in each reaction form different bands on the gel. The bands are used along with reaction labels to read off the DNA base sequence.

However, even with DNA sequencing technology, finding and sequencing a single gene could take many years and, at best, only allow scientists to locate genes to within a million or so bases. So they began wondering what the process would be like if, instead of mapping genes using a few thousand landmarks across the genome, they could have the precise location and order of every single DNA base that existed within each human chromosome. It would be the equivalent of not just knowing a few landmarks around a city, but of knowing every house on every street of that city—in other words, of having a detailed map.

This idea, built on Sanger's sequencing method, would ultimately prompt the Department of Energy and the National Institutes of Health to undertake the daunting task of sequencing the human genome. Today, we look back and realize that completing the genome was a great accomplishment. But its completion also represented a starting point for other important scientific endeavors—for exploring the spectrum of life on earth, for understanding biological processes, and for uncovering how evolution and our biology make us who we are. And to answer these questions we must continue to look back to visionaries such as Mendel, Darwin, Morgan, and others, whose pioneering work laid the intellectual foundation for today's genome studies.

CHAPTER 1 NOTES

1. Although viruses possess many features of life, they are not made up of cells and are not considered to be alive.

2. *Archaea*, once classified as an unusual group of bacteria (*archaebacteria*), are now recognized as a distinct form of life with a unique biochemistry and evolutionary history. *Archaea* possess many genes and several metabolic pathways—most notably those involved in transcription and translation—that are more closely related to those of eukaryotes than to bacteria. This and other evidence suggest that *archaea* might be our prokaryotic ancestors.

3. If all the chromosomes in a human cell were unwound and stretched end to end, they would span about 2 meters (6.5 feet) long, but only about 2 billionths of a meter wide (about one ten-thousandth the width of a human hair). And if DNA floated freely in the cell without the structural support of proteins like histones, it would be in a large knot so fragile that its genetic information would easily break apart, making protein manufacture or DNA replication during cell division difficult. Wound up into chromosomes, DNA is stabilized and its information organized. (You could think of proteins as the packaging and DNA as the content of the package.) During the final stages of cell division, DNA becomes further condensed, so that an entire chromosome is only a few ten-thousandths of a centimeter long.

4. The Indian muntjac, a small Asian deer, is the mammal with the lowest recorded chromosome number and a peculiar mechanism for determining gender: The female has only six chromosomes (three pairs), while the male has seven.

5. Actually, chromosome 21 is only slightly smaller than chromosome 22; it was only as the human genome was being mapped that 21 moved below 22 in the measure of chromosome size.

6. Watson and Crick shared the Nobel Prize for describing the structure of DNA with their competitor Maurice Wilkins of King's College London. But Rosalind Franklin is widely acknowledged as a forgotten codiscoverer. For months, she worked in X-ray crystallography with Wilkins to produce a photograph of the X-ray scattering patterns made from purified DNA molecules. It has been purported that Wilkins showed this image to Watson without Franklin's permission. Watson, possibly drawing on work in X-ray crystallography of helical proteins by Linus Pauling, was inspired by what he saw. Returning to Cambridge, he and Crick quickly set about building their helical model of DNA.

7. Controversy exists over what constitutes a gene. Common to nearly all definitions is that they are units of heredity. They are passed from one generation to the next and convey a phenotype, and so are associated with traits. Since proteins are the primary building blocks of cells, genes are often considered to be regions of DNA that encode proteins.

CHAPTER 2

Of Mice and Men, and Other Creatures

Many people talk about the mapping and sequencing of the human genome as if it occurred in isolation of other genomic research. In truth, the enormous technical challenges the effort presented, coupled with existing interest in exploring the genomes of other more easily studied organisms, motivated scientists at the time to sequence more than just humans. And although interesting in their own right, organisms with much smaller genomes than ours were useful for developing and testing the technologies, laboratory techniques, and computational tools that would make it possible to tackle the human genome.

To understand why working with smaller genomes was critical to the success of the HGP, it helps to get a sense of what the three billion base pairs of our genome represent in terms of information. For example, if you had to type out all the adenine (A), cytosine (C), guanine (G), and thymine (T) bases in our genome as letters, they would fill about 200 Manhattan phone books at 1,000 pages each. But that is only half the challenge. You would also have to put this extraordinary volume of data in sequential order. So it becomes easy to see why scientists initially focused on smaller genomes. After all, if they could not assemble the sequence of a genome a few million bases in length, how would they manage a genome nearly a thousand times larger?

In this chapter we will explore, as did the HGP scientists, a few smaller, more compact genomes that nevertheless represented important sequencing milestones, and in the process learn about the technologies and approaches that led to the completion of the human genome.

BACTERIA

Despite their relative simplicity, bacteria share essential features with much of life on earth. As model organisms, they have been indispensable in the evolution of modern molecular biology as a science, as well as in helping scientists develop and refine the sequencing process.

ESCHERICHIA COLI
(4.6 MILLION BASE PAIRS, OR 300 PHONE-BOOK PAGES)

Escherichia coli, or *E. coli*, is a bacterium that many of us associate with food poisoning. The fact is, most strains of *E. coli* are harmless. The bacterium inhabits the lower intestines of mammals and other warm-blooded animals, where it and other bacteria aid in food digestion and nutrient absorption and production.

E. coli is one of the best-studied prokaryotic model organisms. The bacterium is ubiquitous and surprisingly hearty, making it a reliable workhorse in the laboratory. The K-12 strain of *E. coli* is the one most commonly used in biochemistry, genetics, and physiology. In the biotechnology sector, it serves as the primary "factory" for making many synthetic human hormones, such as insulin.

Since the *E. coli* genome is only a fraction in size of that of humans, it seemed a great initial target for genome sequencing. But because it was the first bacterium begun, project scientists ran into a number of technical challenges. As luck would have it, those working diligently on *E. coli* and learning from their mistakes would have to watch another team grab the honor of being the first to sequence a bacterial genome.

HAEMOPHILUS INFLUENZAE
(1.8 MILLION BASE PAIRS, OR 120 PHONE-BOOK PAGES)

In 1995, while other scientists were wrestling with a "big" genome, a team of scientists led by Claire Fraser and J. Craig Venter from The Institute for Genomic Research (TIGR) in Rockville, Maryland, decided to take the best of the developing sequencing methods and apply them to a smaller bacterium, *Haemophilus influenzae*.

H. influenzae was discovered by German physician and bacteriologist Richard Pfeiffer in 1892 during an influenza epidemic. It was considered the cause of influenza until 1933, when the real cause, a virus, was discovered. Regardless, *H. influenzae* does cause disease, including bacterial meningitis and middle-ear infections in children and pneumonia in children and adults, especially for those with lowered resistance due to diseases such as acquired immune deficiency syndrome (AIDS)—or influenza. Medical importance aside, *H. influenzae* was a sequencing choice because Nobel laureate Hamilton Smith at the nearby Johns Hopkins University could supply high-quality DNA to the TIGR team.

The genome of *H. influenzae* is contained in a single, circular chromosome (as with many other bacteria). To sequence it, the team employed a novel "whole-genome shotgun" sequencing method, in which the bacterium's DNA was sheared into fragments, each about 1,000 to 2,000 base pairs long. From these, about 460 bases of sequence were read from each segment.[1] More than 24,000 DNA segments were sequenced for a total of about 11.6 million bases of DNA, or a greater than six-fold coverage of the bacterium's entire genome.

Why read an extra nearly 10 million bases when the genome size was much, much smaller? The answer has to do with the need to reassemble the genome sequence. Since the DNA segments were randomly generated, their order could not be predetermined, so reassembling the genome would require the use of many redundant, overlapping fragments. Imagine tearing the pages of a book into thousands of fragments. Reassembling the book would require comparing pieces to find edges that matched up. Eventually, you would succeed, but if you had several extra copies of the same book with their pages also randomly torn up, you could use those extra pieces to look for words or phrases that are shared among pieces, and thus more easily reconstruct the entire book.

Redundancy facilitated the reassembly of *H. influenzae's* genome, as well as helped ensure accuracy. Any measurement is subject to error, and highly redundant sequence data allowed scientists to make majority-rule decisions regarding most errors occurring in the sequencing of individual fragments. In the end, by using new computer software programs and a farm of high-powered computers, the *H. influenzae* genome sequence was reassembled into a single closed circle 1,830,138 DNA base pairs long.

Scientists had completed the first sequencing of a free-living organism.[2] Now it was time to search for the genes that encoded the organism's proteins and to determine their functions. Again, the team relied on newly developed, specialized computer programs.

Fortunately, it is easier to find genes in bacteria than in humans. In higher organisms, individual genes are often separated into smaller pieces strung along a length of DNA with intervening stretches of "noncoding" DNA. In bacteria, however, a single section of DNA that encodes a protein is almost certainly an entire gene. Despite this simplification, it was painstaking work to identify the more than 1,700 genes contained within the genome of *H. influenzae*.

Even more difficult was determining what the genes did because that required linking a DNA sequence to the protein it encoded, and then linking that protein back to a likely function. Fortunately, the scientists had an important head start for some of these genes. You will remember from the previous chapter that genes and their functions are highly conserved. For example, genes involved in sugar digestion and metabolism are universal across many organisms. This conservation makes certain genes recognizable to scientists.

By comparing gene sequences in *H. influenzae* to genes whose functions were known in other species, scientists at TIGR were able to assign functions to more than 1,000 (nearly 60 percent) of the genes of *H. influenzae*. Among the hundreds of genes that could not be identified, however, were some that helped differentiate *H. influenzae* from other species of life.

One important medical discovery made possible by the sequencing of the bacterium was a vaccine that has greatly reduced the incidence of bacterial meningitis and ear infections in children.

Following *H. influenzae*, many other fascinating microbes were chosen for sequencing, in part for their uniqueness. *Mycoplasma genitalium*, a urinary and respiratory-tract pathogen, has one of the smallest genomes of any free-living organisms (582,970 base pairs and only 470 protein-coding genes). Even more intriguing is *Methanocaldococcus jannaschii* (formerly *Methanococcus jannaschii*), an extremophile discovered at the base of a deep-sea hydrothermal vent in the East Pacific Ocean, where temperatures approach the boiling point of water and the pressure is greater than 250 atmospheres. *M. jannaschii* was the first member of the *archaea* group of prokaryotes to be sequenced, establishing *archaea* as being distinct from bacteria. Its genome encodes 1,795 genes.

The invaluable lessons learned and techniques employed during the sequencing of these and additional microbial organisms allowed scientists to progress to larger genomes and, in 1997, to complete *E. coli*, allowing the identification of its 4,288 genes.

BREWER'S YEAST

SACCHAROMYCES CEREVISIAE
(13 MILLION BASE PAIRS, OR 867 PHONE-BOOK PAGES)

Another early target for genome sequencing was *Saccharomyces cerevisiae*, commonly known as brewer's yeast (or baker's yeast). *S. cerevisiae* is a versatile organism. Not only has it been used in beer, bread, and wine fermentation for thousands of years, it too is a very important model organism in biological research.

Studying bacteria was essential for understanding DNA's role in managing genetic information, but the lack of certain elements in prokaryotes, such as a nucleus or mitochondria, makes them less than ideal for studying processes that occur in human cells. Yeast, on the other hand, is a relatively primitive eukaryote. It has a nucleus and mitochondria and is a wonderful model for studying a wide range of human diseases.[3] Many proteins that are important in human biology were discovered by studying yeast versions of them, including proteins that regulate cell division, proteins involved in how cells receive signals from other cells and the environment, as well as *enzymes* (which are also proteins) that play a role in how other proteins within the cell are broken down and modified.

Yeast has a larger genome than the first sequenced bacterial species, with its 16 chromosomes, the largest being about the size of the *H. influenzae* genome and the smallest about half the size of that of *M. genitalium*. An international consortium of scientists from the United States, Europe, and Japan took on the yeast genome. Teams met, claimed individual chromosomes, and then worked to complete them, often ensuring the quality of their work by having multiple groups sequence selected regions.

It took five years to sequence the chromosomes of the brewer's yeast genome, which was completed in 1997. But the knowledge gained has led to tremendous insights on many fundamental processes occurring in human cells, thanks in part to the relative ease of finding the yeast *homologues* of human genes (genes that have nearly identical sequences and likely carry out the same functions).

To determine their function and importance, scientists remove certain genes from the genome and observe what happens. "Knocking out" some essential genes in yeast causes it to die, while removing others causes cells to grow irregularly, fail to divide properly, or become unable to use certain food sources. In short, by applying our knowledge of the yeast genome, we have learned about many human cellular and metabolic disorders.

The brewer's yeast genome sequence has also facilitated the sequencing of other fungal species (yes, yeast is a fungus), including many that cause human disease. It has fostered the early development of antifungal drugs that target genes unique to yeasts, including important pathogenic yeasts, as well as the more familiar athlete's foot fungus.

ROUNDWORM

CAENORHABDITIS ELEGANS
(100 MILLION BASE PAIRS, OR 6.6 PHONE BOOKS)

Although yeast is a great model organism because its genome encodes many proteins similar to those found in humans, it lacks an essential element of advanced life: multiple cell types. Yeast has only one cell type, whereas in humans, a single fertilized egg eventually differentiates into approximately 250 primary cell types. To gain insight, scientists turned to a remarkable multicellular model organism.

Caenorhabditis elegans is a transparent nematode (roundworm) about 1 millimeter (0.04 inches) long. Although this "worm" has only about 1,000 cells in its body (we have trillions), like humans it possesses a number of different organs, each of which carries out a specialized function. *C. elegans* is also one of the least complex organisms to possess a nervous system, adding to its value as a model.

In 1974, biologist and Nobel laureate John Sulston mapped the lineage of each of the cells in an adult nematode all the way back to the fertilized egg. By doing so, he had created an excellent microscopic "laboratory" for the study of development and cell differentiation. Sulston also argued successfully for the sequencing of *C. elegans* to demonstrate the feasibility of large-scale sequencing.

In 1998, *C. elegans* became the first multicellular organism to have its genome completely sequenced. The work, completed at the Sanger Institute in Cambridgeshire, England, and at Washington University in St. Louis, Missouri, led to the discovery of nearly 20,000 genes on the worm's six chromosomes. But what distinguished the sequencing was the "divide-and-conquer" approach taken. In the smaller bacterial and yeast genomes, entire chromosomes were fragmented and sequenced using the shotgun method. The genome of *C. elegans* was much larger, so it was initially chopped into segments 35,000 bases long.

Approximately 17,000 of them were mapped along the genome, and a set of about 3,000 segments that covered the genome with minimal overlap was selected. These minimally overlapping segments were then individually sequenced using the shotgun method, after which the genome was "stitched together" using the sequences of these smaller pieces.

As with yeast, sequencing the *C. elegans* genome greatly accelerated research. One particularly interesting cellular phenomenon studied extensively is *apoptosis*, or programmed cell death. Understanding how apoptosis works is essential for understanding the nature of many human diseases, including cancer and type 2 diabetes. In the former, cells escape apoptosis and become "immortal" despite having extensive DNA damage; in the latter, apoptosis inappropriately kills off insulin-producing cells as the disease develops and progresses.

FRUIT FLY

DROSOPHILA MELANOGASTER
(120 MILLION BASE PAIRS, OR 8 PHONE BOOKS)

As discussed in chapter 1, the humble fruit fly, *Drosophila melanogaster*, provided the first evidence that genetic traits were carried on chromosomes. Following Thomas Hunt Morgan's early work, scientists continued to use the fruit fly extensively in genetic studies. Consequently, it became an early target for genome sequencing.

Yet despite the huge community of scientists working on sequencing the fruit fly (or perhaps because of it), the project fell far behind the sequencing of C. *elegans*. It would take genome politics involving the upstart Celera Genomics, along with a bold proposal from one of its computer scientists, Eugene Myers, to put the fruit fly back on center stage.

Myers was not thinking about D. *Melanogaster* itself, but about how the human genome could be more efficiently sequenced. He calculated that it could be decoded using the whole-genome shotgun sequencing method alone, an approach most other scientists believed was all but impossible. Myers's analysis showed that with a sophisticated computer "assembler" program and enough computer power, the genome could be broken up all at once into small fragments, each fragment sequenced, and the entire collection reassembled without the type of intermediate mapping used in C. *elegans* by HGP scientists. But the proposed approach had to first be demonstrated on a smaller, more tractable genome—and the fruit fly genome just happened to be available.

Gerald Rubin of the University of California, Berkeley, the leader of the public fruit fly genome project, was frustrated by the slow pace of its sequencing, particularly given the completion of the worm genome. So when Celera offered to sequence the genome, Rubin and his colleagues jumped at the chance to partner with the company. The group agreed to provide Celera with DNA and other materials, check the sequence, find genes, and assign those genes preliminary functions. In return, Celera would make the data available to the scientific community.

In early 2000, the completion of the D. *melanogaster* genome sequence was announced in the journal *Science*. The fruit fly genome contained about 20 percent more DNA than that of C. *elegans*, but only about 14,000 genes compared to the worm's nearly 20,000.[4] Regardless, the catalogs of genes in both species represented nearly the same diversity of cellular and organismal function.

However, even though the *C. elegans* genome contains more genes, the greater complexity of the fruit fly's physiology and behavior make it a much better model for investigating a wide range of human diseases, including Down syndrome, autism, and heart disease. In fact, a study by scientists at the University of California, San Diego, found 548 fruit fly genes that had human homologues, and more than 700 human diseases were found to have associations with these genes.

HUMAN

HOMO SAPIENS
(3 BILLION BASE PAIRS, OR 200 PHONE BOOKS)

Although the Human Genome Project began as a U.S.-based initiative, it quickly grew to include international participants working together on various aspects of the project. Since everyone needed to coordinate their efforts to avoid redundancy and prevent unfinished gaps in the genome, meetings were held during which groups took responsibility for different chromosomes or chromosome parts. The actual sequencing would be done by the International Human Genome Sequencing Consortium (IHGSC), a group consisting of scientists from 20 institutions representing six countries: France, Germany, Japan, China, the United Kingdom, and the United States.

The public HGP strategy for sequencing the human genome

relied on the same divide-and-conquer approach that had worked for *C. elegans*. This time, however, the genome segments created would be larger—about 150,000 base pairs long. Dr. Pieter J. de Jong, then at the Roswell Park Cancer Institute in Buffalo, New York, took charge of this phase, creating and distributing libraries of *BAC clones*. BACs, or *bacterial artificial chromosomes*, allow DNA fragments to be stably maintained in bacteria, so that scientists can work more easily with them. Hundreds of thousands of unique BACs were mapped to chromosomes; out of them, 35,000 that covered the entire genome were selected for sequencing and reassembly.

During the sequencing, the directors of the major laboratories leading the effort, along with representatives of the funding governmental and philanthropic agencies, met to assess progress and set policy. One such meeting held in Bermuda in February 1996 proved pivotal. Agreeing that the human genome sequence was everyone's property, the participants unanimously committed to making all the sequence information "freely available and in the public domain in order to encourage research and development and to maximize its benefit to society."

This was an extraordinary departure from how scientists usually handle their data. Scientists are judged on the value of their research as demonstrated by publication in scientific journals. Although journals generally require the release of final analyzed data, preliminary results are usually closely guarded from competitors. But scientists in the public genome project felt something more important was at stake: the possibility that even preliminary data could be used to study human disease.

During the 1990s, the HGP scientists made slow and steady progress, developing new technologies and creating "low-resolution" maps of the human genome that would serve as the framework for later DNA sequencing. Yet some scientists began to question whether the project was moving quickly enough. Ultimately, it would be an unusual partnership forged between one of these critics, J. Craig Venter, and a company called Applied Biosystems, Inc. (ABI), that would inject a sense of urgency into the HGP effort and spark the race between public and private interests.

A pioneer in biotechnology, ABI had created the first widely successful computerized, automated DNA sequencing machines—the primary instruments being used by the HGP consortium. Because the company dominated the market, it had also assumed a level of complacency, making only incremental improvements to its machines over time. Then came Molecular Dynamics, a manufacturer of scientific instruments that made its debut in the DNA sequencing market by introducing a somewhat revolutionary instrument that greatly increased the speed at which sequence data could be generated. HGP scientists, particularly those in the large genome centers, began investigating whether these new instruments could replace their hundreds of ABI-brand DNA sequencers.

In response, scientists and engineers at ABI scrambled to create a competing machine. Partnering with the Japanese company Hitachi and using technology developed by Norman Dovichi while at Canada's University of Alberta in Edmonton, ABI introduced its PRISM 3700 automated sequencer in 1998. Soon after, in what some now see as a brilliant marketing move,

ABI announced the creation of Celera Genomics, a company that planned to sequence the human genome in record time using ABI's new instruments. Facing what they saw as a threat, the publicly funded genome centers quickly bought up hundreds of the 3700s to stay competitive with the private company that had now stepped into the race.

Celera's scientists declared they would build on their success with the fruit fly genome by abandoning the HGP's "map-then-sequence" BAC strategy in favor of a whole-genome shotgun approach, claiming it would allow them to finish faster. Whether or not this would have been true, it was a scientific and technical decision that leveraged advances in computation made after the HGP began its work.

Celera's different sequencing approach was not a problem for the HGP, since many scientific advances are driven by competition between groups.[5] But Celera's plans for the genome sequence became a significant issue. Rather than freely release the data, Celera planned to license it, but only after first filing patents on potentially interesting discoveries.

HGP scientists agreed the stakes were too high not to finish the project quickly to ensure the data would be in the public domain. What followed was a greater level of cooperation, collaboration, and coordination among the laboratories, as well as accelerated funding by government and private agencies. They would set aside any differences and focus on generating enough data to cover the genome. It then would not matter much if Celera covered the genome in greater depth by adding HGP's public data to its own.

After a war of press releases, testimony before Congress (where Celera argued that private industry should be doing the sequencing and thus save taxpayer money), and behind-the-scenes mudslinging that threatened both public and private efforts, the race drew to a negotiated end marked by President Clinton's June 2000 announcement and worldwide press conferences.

Another eight months would pass, however, before either group could publish its results, and this too was carefully negotiated. The IHGSC published its results in the February 15, 2001, issue of the journal *Nature;* Celera Genomics published its results in the February 16, 2001, issue of *Science.*

Although the public project's data, including the assembled sequence and the gene count prediction, was made available to everyone, *Science* broke with standard protocol and let Celera publish its analysis without releasing data. As a compromise, Celera would make the data available to those who registered for a weekly download allowance of one million base pairs of data, which could not be shared with others. Unfortunately, it would take nearly 60 years to download the entire sequence at that rate, hardly anyone's definition of "availability."

Although both public and private groups declared that theirs was the better version of the human genome, neither group was in fact finished with the work. Both published papers described a "draft" genome sequence that contained many gaps and errors, and the number of genes discovered, along with their structure and functions, continued to fluctuate as the genome sequence and its annotation evolved.

And so, announcements of completion continued. At an

annual meeting of scientists at the Cold Spring Harbor Laboratory (CSHL) on Long Island, New York, in April 2003, both groups declared the actual completion of the human genome sequence, and the timing was not accidental. First, it fulfilled promises made in the announcement of the draft genome sequence in 2000. Second, the meeting took place shortly before the fiftieth anniversary of Watson and Crick's landmark paper on the structure of DNA—and James Watson happened to be CSHL's president.

It was not until October 2004, two months shy of the original 2005 completion goal, that the public genome project published a paper in *Nature* that described "finishing the euchromatic sequence of the human genome," allowing yet another "victory" to be claimed, despite the fact that additional papers would be published on the completion of 11 of the human genome's 23 chromosomes, including the X chromosome and chromosomes 1, 2, 3, and 4. So somehow the genome was "finished" with only slightly more than half the chromosomes declared completed and the largest chromosomes still far from done.

Regardless of the caveats raised here and the rolling political definition of the word "finished," the production of the genome sequence remains a tremendous technological and scientific achievement. Perhaps the biggest surprise it presented was the relatively small number of genes found—about 25,000.[6]

Astonishingly, only about 2 percent of the sequence encodes for proteins. The remainder consists of noncoding regions, the functions of which might include providing chromosomal structural integrity and regulating where, when, and in what quantity proteins are made.

THE NEXT WAVE OF HGPs

Now that it has begun, human genome sequencing is unlikely to end, and there are many new and exciting sequencing efforts being undertaken. Initiated by George Church of Harvard Medical School in Boston, Massachusetts, the Personal Genome Project aims to initially sequence the genomes of ten individuals and freely release the resulting data and detailed health information. The goal is to eventually enroll 100,000 people and dramatically decrease the cost of human genome sequencing. Launched in January 2008, the 1000 Genomes Project is an international research consortium to sequence 2,500 genomes and create the most detailed map of the human genome available for identifying biomedically relevant DNA variations and their associations with various traits, including many diseases. And a project known as The Cancer Genome Atlas (TCGA) seeks to sequence specific regions of DNA and important mutations associated with various cancers.

A GENOME GRAB BAG

Although the focus of the Human Genome Project was on completing a reference sequence for humans, the scientific and technical bounty it yielded excited the rest of the research world. Scientists working with other species, including plants, animals, fungi, bacteria, and viruses—all with potential applications relevant to human life and health—were motivated to start genome projects of their own.

PLANTS: GENOME GIANTS

The study of plants has been accelerated in many important ways through genome projects. The great thing about plants is that if we discover a gene that improves plant survival during drought, improves yield per acre, or increases vitamin A content, we can create *transgenic* plants by introducing genes that are linked to desired traits into plants, either artificially or through selective breeding.

Although the genomes of our most important agricultural crops—maize (corn), rice, wheat, and soybeans—have been or will soon be sequenced, the first plant genome sequence completed was of *Arabidopsis thaliana* in 2000. This small relative of mustard and broccoli, basically a weed, has a very small genome for a plant at only about 120 million base pairs, roughly equivalent to that of the fruit fly. However, the *A. thaliana* genome encodes nearly 26,000 genes, a number larger than the best-guess number for humans. Scientists chose *A. thaliana* because despite its small genome size, its genes are similar to those of other plants, making it a good model. With its life cycle of only six weeks, the plant's gene functions could be quickly analyzed and the results applied to other plants.

In 2005, the genome of rice, the single most important grain for humans worldwide, was completed, revealing 360,000 base pairs and more than 37,000 genes. The poplar tree, an important model for forestry, was sequenced a year later. Its 550,000 base-pair genome contains more than 45,000 genes. And the 490,000 base pairs of the wine-grape genome were sequenced by a French-Italian consortium in 2007, which discovered more than 30,000 genes.

Why do plants have more genes than do humans or other animals? The answer has to do with their rootedness. If an environment changes, animals can relocate to a place more suitable for their needs. Most plants cannot, so they need larger gene sets, along with the proteins that those genes encode, to adapt to environmental alterations and protect themselves from unavoidable dangers.

Although advances coming out of plant sequencing will allow us to do important things, such as improve the global food supply by creating more resilient, higher-yield crops, plants are not generally useful for medical research. The most useful species for this purpose are mammals, especially the mouse and rat.

MISCELLANEOUS MAMMALS

Despite their small size, mice and rats have genomes containing about the same size and number of genes as our own. More amazing is that, base by base, the similarity between our genes and theirs averages about 85 percent (they are about 88 percent identical to each other). This partly explains why studying human disease in laboratory animals is so valuable. In fact, the mouse has been studied by geneticists for almost as long as the fruit fly, and many genes in humans were discovered by first finding the corresponding genes in the mouse. Being slightly physically larger, the rat is a better model than the mouse for studying physiology, and its use has led to discoveries in heart disease, diabetes, and other disorders.

Scientists have also sequenced the domestic cat and dog, again finding genomes comparable to ours. The information collected may one day help us decipher biological characteristics of higher organisms that go beyond physical traits and involve behavior. Why, for example, do certain dog breeds exhibit behaviors that are, for lack of a better word, inbred?

And the negligible differences among human racial groups are put in perspective when considering how a Chihuahua and a Great Dane have nearly identical genomes, yet quite disparate phenotypes. There must be some very important genes in dog genomes wherein small DNA base changes can produce such significantly different characteristics.

Agriculturally important animals like the cow, sheep, and pig are also being sequenced, as are the horse, brown bat, rabbit, and platypus. The list goes on, and what we are repeatedly seeing is that among mammalian species, the differences found in the DNA of their genes tend to be relatively small.

To determine how these small differences help separate humans from other species, scientists are examining our closest relatives, including the chimpanzee, gorilla, and Neanderthal. Surprisingly (or not), the genome of the chimpanzee, our closest living relative, is more than 98 percent identical to our own. The Neanderthal genome, derived from DNA isolated from fossilized bones, is nearing completion, and the data available thus far is confirming that the oft-derided Neanderthals were more like us than we might like to admit.

THE UNSEEN WORLD

Scientists will also continue to sequence countless bacteria, viruses, and other disease-causing organisms, and genome technology is poised to become a first-line tool for uncovering the interplay between human hosts and the microbes that we all carry. Just as we have learned that the human papilloma virus (HPV) can lead to cervical cancer and the bacteria *Helicobacter pylori* causes ulcers, we are now using genome sequencing to search for microbial causes for diseases that we have not, until now, considered as having an infectious component.

CHAPTER 2 NOTES

1. The DNA fragments were first "cloned" by biochemically inserting them into a small, circular piece of bacterial DNA called a *plasmid*. Each plasmid was then introduced into an *E. coli* bacterium that did the work of making millions of copies of the cloned DNA segments to be sequenced using Frederick Sanger's method. The 460 base reads were at the limit of what was possible from a single sequence read in 1995.

2. The genomes sequenced before *H. influenzae* were viral genomes only a few thousand bases long. It was Frederick Sanger who, in 1977, sequenced the first genome: the virus bacteriophage □X174. Because *H. influenzae* was a "free-living" organism, it had a much larger genome. Also notable is that Sanger invented shotgun sequencing when he sequenced the genome of bacteriophage λ in 1982.

3. Scientists do not always spend their lives in labs. Some early brewpubs in the United States were founded by biologists who could culture yeast. In fact, most large breweries run yeast biology research programs, and brewery scientists have contributed resources to the broader scientific community. One yeast strain used in an early part of the HGP was AB1380, where "AB" stands for Anheuser-Busch, the brewers of the American Budweiser and the provider of that yeast strain.

4. An organism's size does not necessarily relate to its genome size. And genome size, at least for eukaryotes, is not necessarily related to the number of genes encoded. The single-celled *Amoeba proteus* has 290 billion base pairs in its genome, and the related *Polychaos dubium* (formerly *Amoeba dubia*) has 670 billion base pairs—223 times that of the human genome.

5. Celera Genomics did use public HGP data that included all of the mapping information and landmarks that the public BAC data provided. The company later claimed it did not need the extra information.

6. The absolute number is still in question, but the tally is generally acknowledged to be between 22,000 and 33,000 genes. Almost any number can be contested, however, since most genes have multiple forms (an average of five) that may or may not encode slightly different proteins. Depending on how precisely you define a gene, the number could be 22,000 or 110,000. For now, the consensus in the scientific community is about 25,000.

CHAPTER 3

Reading the Book of Life

The first question almost everyone asked about the completed reference human genome sequence was, "Whose genome was sequenced?"

The answer is complex and differs for the public and private genome projects. From the start, the public project was charged with examining the ethical, legal, and social issues associated with genomics and its applications, and one of the driving principles guiding it was the need to protect the rights and privacy of any individual whose genome was being mapped. This is a standard practice in biomedical research, where the most valuable resources for studying disease are human samples, yet where, unfortunately, stigma is sometimes attached to the disease being studied.

Since a person's genome represents highly personal information, scientists and others worried about its misuse. For example, a genome sequence could infer paternity or other genealogical features.[1] Variants found in a sequence could be used to claim statistical evidence about one's risk for diseases, which can affect anything from personal relationships to finding a job or becoming insured. A genome sequence could reveal the potential for diseases lacking effective therapies, raising the associated risk for psychological stress. These and other considerations greatly motivated the project organizers to protect the identities of those people whose DNA was eventually sequenced and to ensure that the data could not be linked back to any individual.

In biological studies conducted in most of the world, participants must provide informed consent before their tissue or other samples (or even demographic information) can be used. They must sign a document outlining the purpose of the study, its necessity, and any risks associated with participation. This posed a challenge regarding the HGP because it was unclear to many how much information would be collected and how it could, and would, be used. The scientists who understood the project goals and risks could have conceivably donated their DNA; however, biomedical studies cannot focus on one gender, ethnic, or demographic group without very good reason. So newspaper ads were placed seeking volunteers who, after being educated about the project, provided consent and donated blood or sperm samples that were subsequently rendered unidentifiable. Four samples, taken from two men and two women, were randomly selected for the project.

International research groups that collected additional samples representing their respective populations used similar sample collection methods. Yet despite randomization efforts, it is known in the genomics community that the primary donor for the HGP was a male, code name RP11, from the Buffalo, New York, area.

Celera Genomics began with DNA samples from 21 people, out of which it selected five for sequencing. One sample provided the bulk of the data, while the other four were studied for variations that would later prove useful for studies involving disease mapping. Although the samples were collected under strict ethical guidelines and supposedly treated equally, one of them was not quite anonymous. After the announced completion of the draft genome, J. Craig Venter confirmed rumors in the scientific community that his DNA sample was among the five. And although he never admitted it, Venter strongly hinted that the Celera genome sequence was largely his own.

Although most of the public HGP data came from a single unknown person and much of the Celera genome sequence was admittedly Venter's, the human genome sequence belongs to everyone. With more and more data accumulating, we now have concrete scientific evidence that the differences between any two people is about one-tenth of 1 percent, regardless of race, ethnic background, or other physical characteristics, making the reference genome sequence produced by the HGP an excellent representation of us all.

TAKE A LOOK AT YOURSELF

For anyone interested in viewing the public genome sequence, three sources make it freely available to the scientific research community and the broader public. The first is the National Center for Biotechnology Information (NCBI) of the National Library of Medicine at the NIH in Bethesda, Maryland.[2] There, the human genome sequence is maintained as part of a larger database called GenBank and is integrated with other resources that link the genes and sequence data to published scientific papers, gene and disease databases, information about human variation, and data and information on similar genes in hundreds of other species.

The European Bioinformatics Institute (EBI) in the United Kingdom has a similar database housing the genome sequences of humans and a large number of other species.[3] The Ensembl database identifies and describes a set of genes within the human genome that differs from the set at NCBI—not surprising, given the difficulty in defining genes.

Finally, there is a widely used database at the University of California, Santa Cruz (UCSC).[4] As the public HGP was nearing the end of its race against Celera, scientists who had been focused on generating sequence data realized they had to also assemble the genome from the data. David Haussler and Jim Kent of UCSC stepped up to the challenge, and their results were presented in the UCSC Genome Browser. This source is a favorite of scientists because it is annotated with other information and sources of data.

Although the availability of sequence data might inspire you to view the genome, these sources are not very user-friendly. Each has some type of tutorial to help users sort through the information but, in truth, the sites are designed by scientists for scientists. Practically speaking, it is hard to anticipate the questions millions of people might ask about the data and design a Web portal with answers for casual visitors. In addition, the data are so large and complex, and what we know about the genes and the genome so preliminary, it is difficult to present it all in a useful, non-misleading way. Nevertheless, if you read an article about a gene being linked to a disease and want more information on it, the sources provide helpful Web links for the nonscientist.

WHAT'S IN A NUMBER?

Before the HGP began, estimates of the gene content in the human genome ranged from about 40,000 to more than 100,000.[5] With sequence data now available and three major scientific groups providing different views of it, more precise predictions could finally be made.

Nevertheless, identifying genes in complex eukaryotic genomes is difficult. In prokaryotes like bacteria, protein-coding genes are linear, contiguous stretches of DNA bases. But in eukaryotes, like humans, genes are most often segmented into small fragments that contain "functional" protein blueprints called *exons* separated by intervening noncoding sequences called *introns*.

The lengths of exons and introns vary highly; many genes can span 10,000 or more bases, even though their actual coding sequences are typically only 1,000 bases or fewer long. The distribution of genes is also not uniform. Some regions contain a high density of genes and other regions appear to be "gene deserts." These and other factors complicate efforts to find the "real" genes that cells use to make proteins.

Another complexity of gene-counting involves a stage during protein manufacture that occurs after DNA transcription. Before precursor ("immature") mRNA is transported to the ribosomes for translation into protein, it is taken up by a body in the nucleus called the *spliceosome*, which removes the introns and joins the exons together to create a "mature" mRNA molecule. During this process, the mRNA can be spliced together in slightly different ways, and these splice variants might or might not encode exactly the same proteins. Scientists estimate that genes have, on average, about five protein-coding splice variants. Often, the variant that survives the splicing process depends on the tissue, developmental stage, or disease state of the cell in which the gene is being expressed.

Whatever the considerations are that keep us from nailing down an exact number of genes in the human genome, the current estimate of 25,000 is still far less than many expected when considering the complexity of humans compared with *C. elegans* (the roundworm) or *D. melanogaster* (the fruit fly), both of which have nearly the same number of genes as we do; or with plants, which have gene totals that typically exceed ours.

A partial explanation might be that splice variants are not included in the gene total estimate. Or, perhaps proteins are modified after they are made, something that cells do. However, a more reasonable and widely accepted explanation is that the intricacy of an organism is determined not by the number of proteins its genome encodes, but by the complexity of the networks assembled from those proteins.

Protein-coding sequences are not the only important features encoded in the genome. A recently discovered class of RNAs transcribed from the genome are noncoding RNAs called *microRNAs (miRNAs)*. Although these RNA molecules do not encode proteins, they might play a very important role in the cell, acting as a negative regulator of other genes by preventing their RNAs from being made into proteins. Many different miRNAs also appear to be active or silent in various diseases, including cancer, which makes their role in possibly promoting or preventing disease the subject of intense study.

These and other characteristics of DNA and protein manufacture have us reevaluating how we think about genes—what they are, what they do—and have put the goal of determining an exact gene count in perspective. Whether there are 24,000 or 26,000 or some other number of protein-coding genes is less important than understanding their functions and knowing which gene variants are associated with particular physical traits.

LINKING GENES TO TRAITS AND FUNCTIONS

The original motivation for sequencing the human genome was to accelerate the process of finding genes associated with diseases. When a gene responsible for an illness is discovered, scientists can try to design diagnostic tests, develop strategies for earlier detection, or recommend lifestyle changes for mitigating risk. For example, if you learned that you carry a gene predisposing you to heart disease, you would likely watch your cholesterol and blood pressure more closely than you would if you did not carry the gene.

Finding genes can also make it easier to figure out what causes many diseases. Looking again at heart disease, we might discover mutations in genes that are part of pathways—interacting groups of proteins and cellular metabolites—that govern lipid metabolism (the way the body's cells use fats). By knowing which pathways are affected, we could make specific dietary recommendations for preventing heart disease or develop drugs that could help correct the problems occurring in those pathways, and thus perhaps reduce the risk of heart attack. So, rather than wait for heart disease to strike, a person can take better control of his or her health early on.

Although promising, this type of medical advancement requires finding the gene or genes associated with a disease. The map of the human genome has not solved this problem, but it has made the search easier. In chapter 1, we used an analogy involving Woody Allen to explain the concept of mapping traits to specific regions of DNA. It is worth introducing another analogy here to illustrate the value of a human genome sequence map in the hunt for disease genes.

Imagine that you are in Paris, France, and would like to take a train to Lyon. Paris has six major train stations: Gare d'Austerlitz, Gare de l'Est, Gare de Lyon, Gare Montparnasse, Gare du Nord, and Gare Saint-Lazare. If you were a Parisian, you would know this, as well as know that the train for Lyon leaves from Gare de Lyon. But if you were a tourist who did not speak French and knew nothing of Paris, getting to Lyon could pose a problem. A map would help tremendously, since instead of wandering around the city trying to ask for directions, you could consult your map to locate train stations. If your map included the station names, you could target Gare de Lyon.

Where we are today with the human genome is somewhere between the best- and worst-case scenarios of this analogy of either being a Parisian or a tourist without a map. Now that the genome sequence is complete and we have a map, the goal of biomedical science is to annotate it with gene locations and functions.

So how do we assign a trait or a function to a specific gene? To answer this, we need only look back at Mendel's principles of genetics, which explain how physical traits are passed from one generation to the next through discrete units of heredity transferred from parent to offspring. Although Mendel failed to recognize these units as being (primarily) protein-coding elements encoded within the genome, his idea of examining heredity is still an excellent way to link genes to functions.

Specifically, the easiest way to map a gene to the human genome is by studying families in which a given trait is more common than in the general population. If certain traits are associated with genetic differences—if a gene is tied to a trait—then a change in the gene should be tied to a change in the trait. For example, we know that colon cancer is associated with known mutations in a number of genes, and we suspect we may find many others; but one inherited form of colon cancer is associated with a disease called familial adenomatous polyposis (FAP). Patients with FAP have large numbers, often hundreds, of polyps that form primarily in the large intestine. Left untreated, these polyps can undergo malignant transformation into colon cancer, and they do so with a high degree of regularity.

Studies of families in which FAP was common suggested that the disease was due to a single dominant gene. (If one parent had the mutant gene, the children had a 50 percent chance of inheriting it.) By looking at family members with and without FAP and searching for similarities and differences in their DNA, scientists were able to find a gene now known as adenomatous polyposis coli, or APC. If mutated, the APC gene causes individuals to develop FAP. Scientists are investigating exactly what APC does and are trying to develop a drug to prevent patients with the mutated gene from developing FAP. But merely identifying the gene and the mutation that results in the disease has been extremely useful because individuals at risk can now be monitored for FAP and treated if it develops.

In the case of FAP, the link between the disease and the APC gene appears to be straightforward. Unfortunately, most diseases are not caused by single mutations, but by networks of genes acting together and with the environment (as we will see in later chapters), and so even the human genome sequence does not provide a simple "answer key" for linking genes with traits. But we are better equipped now to use the principles of heredity to find specific genes and to do so more quickly. In fact, between the time at which these words went to press and when you read them, new associations will have been found between genes and diseases and new approaches to therapy devised.

MUTATION: IN THE EYE OF THE BEHOLDER?

In our discussion thus far about the efforts to link genes to human disease, we have often referred to changes that occur in the DNA. Understanding these changes is critical to deciphering the nature of disease itself. Yet the relationship between genes and traits is extremely complex and our understanding of gene variants continues to evolve.

This is partly reflected in the terminology used by the scientific community. We used to refer to normal ("wild type") and mutant forms of genes, but today we generally discuss variations, or *polymorphisms*, in gene sequences. The reason is simple: It is not easy to define what is normal and what is a bad change. Which is "better," blue or brown eyes? Black or blond hair? Any answer is generally arbitrary and, at most, a matter of personal preference. But the diversity of human characteristics, as irrelevant as it might be, arose through genetic changes that occurred during our evolutionary history. Consequently, it is well worth examining the concept of mutation—when it is good, bad, or neutral.

For many, the term "mutation" is loaded with negative connotations, calling to mind science-fiction films about altered humans or giant irradiated insects terrorizing the world. The word "mutant" says it all. But Hollywood's idea of mutation is simplistic. The fact is, there are different types (and causes) of gene mutations. The classifications are extensive but, in general, mutations can be deleterious, advantageous, or neutral when it comes to the ultimate fitness or survival of an organism. Yet even these basic classifications are relative, since genetic mutations known to be potentially detrimental are not always eliminated from the population over time through natural selection. Instead, they are allowed to persist, leading us to ask what constitutes a "normal" gene anyway? Sometimes it all depends on context and circumstance.

Consider one of the most common genetic diseases worldwide, sickle cell disease (or sickle cell anemia), a disorder characterized by red blood cells that assume a rigid, sickle shape instead of their normal disc shape when the body needs oxygen, such as during exertion. These abnormally shaped blood cells become stuck in small arteries, cutting off blood flow to muscles and other organs and generally causing terrible episodes of pain lasting from hours to days. Sickle cell disease carries other complications, including anemia, infections, strokes, and heart and lung problems. Tragically, life expectancy is shortened—about 42 years for men and 48 years for women.

The cause of sickle cell disease is a single mutation in the beta-globin gene, which encodes part of the hemoglobin molecule that red blood cells use to carry oxygen throughout the body. This mutant form of the beta-globin gene is often called the "sickle cell gene" and is *autosomal recessive*, meaning that a person who carries a single copy does not manifest the disease, but is instead *heterozygous* for the trait (or a "carrier") and can pass it on to his or her children. If both parents are carriers, there is a one-in-four chance that their child will get two copies of the mutant beta-globin gene and thus be *homozygous* for the trait and develop the disease.

The sickle cell gene variation and the disease occur more commonly in people (or their descendants) from tropical and subtropical regions where malaria is or was common, including regions of Africa, the Mediterranean Basin, the Arabian Peninsula, and the Indian subcontinent. Their occurrence is most common in sub-Saharan Africa, where as many as one-third of inhabitants carry the mutant version of the gene. Interestingly, if we took a map of these areas and overlaid it with a map of the historical prevalence of malaria, we would see an almost perfect correlation. And therein lies the explanation of the mutation.

Carriers of the mutant sickle cell form of the beta-globin gene are protected against the infection and the most severe neurological disorders associated with *Plasmodium falciparum*, the parasite that causes malaria. In areas where the sickle cell gene arose and became common, two competing selective pressures existed side by side: malaria and sickle cell disease. If malaria was a serious threat at a particular time, you had a better chance of surviving to produce offspring if you carried the sickle cell gene. But if malaria was not that much of a threat, you were better off carrying the normal beta-globin gene instead.

Considering Darwin's natural selection, here is a situation where a gene considered detrimental, and which would therefore be selected against, is in some cases beneficial and selected for. In other words, although sickle cell disease is bad, malaria has historically been a more deadly killer, and so the gene was preserved because it helped people reach reproductive age during certain periods in history.

Countless other gene mutations have their relative benefits.

Humans originated in Africa and had dark skin, as do Africans of today. Skin color is determined by melanin, a pigment everyone possesses to a greater or lesser degree. Large amounts of melanin protect against damaging ultraviolet radiation present in sunlight. But sunlight also allows the human body to synthesize vitamin D, which helps maintain organ systems and is essential to bone growth and development. When people eventually migrated out of Africa into Europe, the sunlight farther north was much less intense, so people's ability to synthesize vitamin D declined. As a result, mutations for lighter skin quickly evolved in early Europeans for more efficient vitamin D synthesis. Yet although this mutation is beneficial for this group of people, it would pose a serious disadvantage for people in Africa.[6]

Eye color is another interesting example of a mutation serving a "helpful" purpose. At least three genes are involved in determining eye color, and the rules for inheritance are complex. But the genes associated with brown eyes are generally dominant over those for blue eyes. So why are blue eyes fairly common? Well, it is an evolutionary advantage for a blue-eyed man to take a blue-eyed woman as his mate. If the pair has blue-eyed children, all is well. But if the woman bears a brown-eyed child, it is a sign the biological father is someone else. And since evolution is all about ensuring that *your* genes live on, discovering that a child is not yours might make you less invested in its survival to adulthood (speaking here, perhaps, for early man).

It is interesting to see how this bit of natural selection trivia echoes on in today's sexual preferences. A study found that, on average, women show no preference for brown- or blue-eyed men irrespective of their own eye color (a woman knows a child is hers) and brown-eyed men also had no eye color preference for women (their children are likely to be dark-eyed). In contrast, blue-eyed men show a strong preference for blue-eyed women.

POPULATION GENOMICS: SOMETHING FOR EVERYONE

The sickle cell gene would likely not have been discovered had scientists only studied a small pool of random individuals, illustrating the value of studying affected populations to uncover disease-causing mutations. In fact, nearly every gene associated with a human trait has been found through genetic studies of populations where the trait is more common, on average, than in the rest of humanity.

For example, BRCA1, a gene important in many forms of breast and ovarian cancers, was identified by following patterns of inheritance in Ashkenazi Jewish families exhibiting a very high incidence of breast cancer. As the descendants of medieval Jewish communities of the Rhineland (in the west of Germany), Ashkenazi Jews were a nearly ideal subpopulation to study because their family structures are well-defined—a characteristic of relatively "isolated" communal groups, in which, for cultural or historical reasons, members tend to marry within their group and family trees are generally well-detailed.

The discovery of the BRCA1 gene (and its more than 600 mutations identified thus far) has implications for all of us. Available tests for the mutations do not consider race or ethnicity and instead yield the information any person needs to manage his or her health care. A woman carrying a mutation in BRCA1 would be advised to at least increase the frequency of breast exams. Some women, particularly those with many cancers occurring in their families, or who have had breast cancer, undergo prophylactic mastectomies. Whatever the decision, it can be made based on genetic rather than ethnic or racial information.

This universality holds true for nearly every mutation or gene discovery. Although certain gene variants are more or less common in some populations, most disease-causing mutations exist in every group to some degree because what we as humans share is far more extensive than the small differences that separate us. Genome sequencing has revealed that we are about 99.9 percent genetically identical to one another. On average, the genomes of any two people differ by only one base out of every thousand, with most differences existing in regions that do not represent genes, but rather the *intergenic* space between them.[7]

Even so, thousands of differences exist between the gene sequences of any two individuals. Most are irrelevant and many do not represent changes in the respective proteins. But in terms of research, this tenth of a percent genetic variation is important because it helps scientists use the human genome sequence to everyone's benefit.

GENE REGULATION AND EXPRESSION

The tale of the Human Genome Project is one of extraordinary success, but it is only part of the ongoing genomics story. The genes within the genome represent a parts list, not a complete blueprint for making a human cell, let alone a human being. The focus for now is on using genome sequence information to learn how genes work together and how the processes associated with them can go wrong to result in disease.

A cell, remember, is basically a "machine" made of proteins, and every cell in the human body carries the same DNA with the same collection of genes. But the body's 250 or so cell types—from skin cells, pancreatic cells, and neurons to kidney cells, cardiac muscle, and lung cells—perform different functions and therefore require different proteins to carry out their jobs. For example, brain cells produce and release neurotransmitters, something liver cells do not need to function. Liver cells, on the other hand, must produce numerous enzymes to execute important tasks such as detoxification and digestion. Since all cell types share the same DNA blueprint, this implies that cells use genes selectively to accomplish their diverse functions. To do so requires that certain genes be "turned on," or expressed, and others be "turned off," or silenced, at any given time.

Understanding this differential *gene expression* and its regulation, as well as how genes work together in networks and pathways, is central to understanding many biological processes, including development and disease. It is all part of an extraordinarily complex yet elegant feedback and control system: Genes encode proteins, the proteins fit together in pathways and networks in ways that allow cells to function, and some of these networks in turn control which genes are turned on or off.

One of the many ways that cells regulate gene expression is through *epigenetic modification* (from the Greek *epi* for "over" or "above"), which involves activating or deactivating genes by making minor biochemical changes in their DNA, or in the proteins that hold chromosomes together, without altering the DNA sequence of the genes. In other words, DNA can be modified in a manner other than mutation. The best-studied type of epigenetic modification is *DNA methylation*, a process that subtly alters a cytosine base by adding to it a methyl group (a hydrocarbon group related to methane and composed of one carbon and three hydrogen atoms with the chemical formula $-CH_3$), which changes the molecular structure of the cytosine without changing its base-pairing properties.

How gene expression is regulated, such as through epigenetic modification, is very important because the manufacture of too little or too much of a particular protein by a cell can lead to problems, including disease. Errors that occur with DNA methylation have been linked to a wide range of diseases, including two developmental disorders: Angelman syndrome and Prader-Willi syndrome. Both diseases are associated with genes that reside on a small region on chromosome 15 (about one million base pairs long).

It so happens that parts of the genome are methylated during the production of gametes (sperm and egg sex cells). Normally, the maternal copy of this region of chromosome 15 is expressed (or unmethylated) and the paternal copy is silenced (methylated). However, if an error occurs so that the region is instead methylated in the egg, the child born can develop Angelman syndrome, a disease characterized by epilepsy, tremors, and a perpetually smiling facial expression. If both sperm and egg copies end up unmethylated, the result is Prader-Willi syndrome, the symptoms of which include hypotonia (abnormally low muscle tone), obesity, and hypogonadism (wherein the testes or ovaries underproduce hormones).

Epigenetic modification, also referred to as "the second genetic code," is the subject of intense study, including two new large-scale projects: the NIH Roadmap Epigenomics Program in the United States and the Human Epigenome Project in Europe. Both intend to examine genome-wide epigenetic alterations and the role they play in changing gene expression patterns involved in development, cellular differentiation, and disease.

CHAPTER 3 NOTES
1. The discovery of mis-paternity is common in genetic studies, where average rates are between 10 percent and 20 percent (and, interestingly, higher for second children). The informed consent process typically specifies that paternity may be revealed as a result of genetic testing performed, and studies must have protocols in place for handling such information.
2. http://www.ncbi.nlm.nih.gov/projects/genome/guide/human/
3. http://www.ensembl.org/Homo_sapiens/Info/Index
4. http://genome.ucsc.edu/cgi-bin/hgTracks?org=human
5. In 2000, a "gene-count sweepstakes" was announced at an annual meeting at Cold Spring Harbor Laboratory, with guesses costing $1 in 2000, $5 in 2001, and $20 afterward. At the 2003 announcement of the final-draft genome sequence, the winnings were split among three scientists with the lowest guesses, all of which were still far higher than the resulting estimated gene count of 21,000 at the time.
6. Protective clothing and sunscreen largely mitigate the problem for light-skinned people living in tropical areas, and supplements help dark-skinned people living in areas with lower levels of sunlight to compensate for diminished vitamin D production.
7. Amazingly, only about 1.5 percent of the genome contains actual protein-coding gene sequences. The rest of the genome, often referred to as "junk DNA," probably plays many important biological roles that have yet to be determined.

Genomics Applied: Tackling Cancer

The hope that genomic data could lead to major advances in understanding, diagnosing, and treating disease led many people to speculate as to how the human genome sequence would be used. It was one thing to generate masses of data, but another to apply that enormous wealth of information toward practical research purposes. Ultimately, what turned hope into reality was not the sequence data itself, but the technological breakthroughs that revolutionized our understanding of biological systems and diseases. These advances allowed the study of biology to evolve from being purely a laboratory science into an information science, for which effectively handling and analyzing tremendous quantities of data is as critical to research success as is the choice of experimental system.

Genomic data analysis involves integrating data with a vast body of existing information on biological systems. Complex datasets must be presented in easily interpretable formats, which is no small feat. To meet these challenges, a new field called *bioinformatics* arose. Combining biology, computer science, and information technology, bioinformatics addresses the need to manage and make sense of the flood of data that genomics has generated. And although bioinformatics scientists played a huge role during the Human Genome Project in assembling genomic sequence data and finding the genes within, the science came into its own as new genomic technologies were applied to the study of human disease.

One of the first diseases to receive significant attention from scientists who were eager to experiment with new sophisticated genomic and computational tools was cancer, an ideal target for many reasons. First, cancer has a heredity component, which suggests that some genes involved in the disease can be discovered by tracing it in families. Second, although many cancers are not hereditary, the genome itself is nearly always altered in cancer, indicating that the disease's development and progression rely on the disruption of certain genes and their functions. Third, unlike most other diseases, cancer is generally treated through removal of the diseased tissue, providing a ready source of patient material for research. Last, and perhaps most important, the National Cancer Institute has the largest research budget of all U.S. National Institutes of Health centers and institutes, meaning that more research dollars are available to study cancer than any other single disease. Consequently, cancer became the most widely studied disease using genomic technologies and has paved the way for investigating numerous other illnesses.

What Is Cancer?

Few people today have not been touched directly or indirectly by this insidious disease. In the United States, cancer is the second leading cause of death (behind cardiovascular disease), striking more than 1.2 million Americans—one diagnosis every 30 seconds. It is the leading cause of death for those between 45 and 65 years of age and of children between the ages of 1 and 14 years. Lung and prostate cancer are the top cancer killers for men; for women lung and breast cancer are most common. One in two men and one in three women will be diagnosed with cancer in their lifetimes, and nearly one in five Americans will die from it.

Contrary to what many people think, cancer is technically not a single disease, but a group of diseases characterized by uncontrolled cellular growth, the invasion of a tumor into adjacent tissues, and the *metastasis*, or spread, of tumor cells to other locations in the body via the lymphatic system or blood circulation. These criteria distinguish cancerous tumors from benign, rarely life-threatening tumors and explain why a failure to control the disease often results in death.

Nearly all cancers are caused by abnormalities that arise in DNA and can be due to the effects of cancer-causing agents (carcinogens) such as tobacco smoke, radiation, chemicals, or viruses. Other genetic changes can occur spontaneously through errors in DNA replication. Inherited genetic variants that prevent a cell from correcting other genetic errors or from self-destructing when errors occur can also contribute to the development of the disease.

Perhaps the most frustrating aspect of cancer is the difficulty in treating it. Although the human immune system has mechanisms for recognizing disease-causing organisms and targeting them for elimination, it is generally unable to launch an attack on tumor cells because they are the body's own cells gone haywire.

Thankfully, during the past 50 years scientists have made great strides in developing cancer treatments that have yielded dramatically higher survival rates for patients. Chemotherapies have improved, both as single drugs and as "cocktails" of drugs that work together to target multiple weaknesses in the cancer, most often by targeting rapidly growing cells and preventing them from dividing into two new cells.[1]

Unfortunately, chemotherapy drugs have their drawbacks. Because they are not perfectly selective, they affect all the cells in the body and can cause intense side effects. In addition, not all tumors respond to therapies and still others become resistant, so the search continues for better treatments, as well as for drugs that target the mechanisms that allow cancer to develop in the first place.

Perhaps the most critical weapon in the fight against cancer, however, is early detection, since preventing a tumor from spreading affords the highest likelihood that the disease can be "cured."[2] In the developed world, screening tests such as mammograms, pap smears, and colonoscopies are invaluable tools for early detection. But tests like these are only the beginning. Researchers hope to go further by developing others that can detect tumor cells circulating in blood or identify the products of their unique metabolism in blood or urine.

LINKING GENES AND CANCER

Human cells are amazing. They grow and divide to keep our bodies intact and allow us to adapt and respond to environmental changes. Under normal conditions, what is known as the *cell cycle* begins with a new cell's growth and its preparation to divide. The cell creates a copy of its DNA and then divides into two daughter cells, marking the beginning of a new cycle.

Throughout the division process, known as *mitosis*, the cell makes numerous checks to ensure that everything is proceeding correctly and that the daughter cells will have accurate copies of the original cell's DNA. If its DNA is damaged beyond repair, the cell will divert its activity to apoptosis—programmed cell death. Cancer develops when genes involved in the cell cycle malfunction. Rather than self-destruct, cells with damaged DNA will grow and divide without restraint, eventually invading neighboring tissues and other parts of the body where secondary tumors (metastases) are established.

Many people think of cancer as being a primarily inherited disease. Although all cancers are genetic in that they are triggered by altered genes, only a small fraction of cancers are actually inherited; that is, caused by mutations passed from generation to generation. Most cancers arise from random mutations that occur within a person's cells during his or her lifetime, either as errors during cell division or from injuries to DNA caused by environmental carcinogens. If mutations strike the right genes, a cancerous tumor can develop and progress.

There are three classes of genes that are typically mutated, separately or together, in cancer. The first are *oncogenes*, which, in their normal state, encourage cell growth. When mutated or "over-expressed," they can flood a cell with signals that tell it to keep dividing.

Tumor-suppressor genes restrain cell growth. If these genes are missing from or inactivated in the cell, uncontrolled growth and division will occur. BRCA1 and BRCA2, the inherited genes that predispose women for breast and ovarian cancer, are examples of malfunctioning tumor-suppressor genes. Many other cancers have also been linked to mutations in these genes, including colon cancer (the APC gene) and retinoblastoma (the RB1 gene). The best-known tumor-suppressor gene is p53, which has been associated with nearly every type of cancer.

Finally, *DNA-repair genes* trigger cancer not by spurring cell growth, but by failing to correct mistakes that occur in DNA during cell division. During its lifetime, a cell can accumulate thousands of mutations in its DNA. If these "hit" critical oncogenes or tumor-suppressor genes, cancer can develop. Most models of cancer development require multiple hits to occur in the cellular replication machinery to spur growth, disable tumor-suppressor genes, and eliminate checks on the integrity of the DNA.

One of the difficulties in treating cancer is that these genetic changes also help tumors adapt and survive. By rapidly growing and accumulating mutations, tumors can elude treatment. Chemotherapy is essentially an evolutionary challenge for a tumor: It will either adapt to the drugs targeting it or perish. Unfortunately, when a tumor does evade chemotherapy, few treatment options remain, and a patient faces a poor prognosis.

To better understand cancer and develop improved therapies, we must learn about the individual genes responsible for the disease's development and progression, the pathways and networks in which they participate, and their ability to survive chemotherapy. Over the years, significant advancements have been made in these areas. But before we can discuss them, we must first examine the genomic technology responsible for changing the face of biomedical research—the *DNA microarray*.

SURVEYING GENE EXPRESSION

In the previous chapter, we talked about how the same DNA blueprint is used to create the different cell types of the human body and how gene regulation is involved in the manufacture of the specific proteins a cell needs to function. The regulation of gene expression applies to cancer cells, too, since certain genes must be turned on or off for a normal cell to become a tumor cell. Knowing which genes are turned on or off in cancer is invaluable for unlocking the mysteries of the disease, and thanks to the work of a British molecular biologist, we have a relatively simple way to track gene expression.

In 1975, Edwin Southern sought a method for determining whether a certain DNA sample he had been working with encoded a mutant form of a gene. Southern knew that in its natural state DNA is a double-stranded molecule, but that it could also be denatured, or separated, into two single-stranded molecules. Once separated, the resulting complementary strands could also hybridize back into a double-stranded molecule using base-pairing rules. Southern realized that if this worked for natural DNA, it could also work for synthetic pieces of DNA. And if synthetic pieces were radioactively labeled, the presence or absence of their complementary sequences could be detected.

This technique, known as the "Southern blot," involves breaking apart a cell's DNA into small fragments using *restriction enzymes*.[3] The fragments are denatured and separated by size using a technique called gel electrophoresis, then bound to a nitrocellulose membrane, producing a "smear" of single-stranded DNA, with the largest fragments appearing at the top and the smallest at the bottom. Synthetic pieces of DNA, with specific base sequences of interest, are then used as probes on the DNA smear. If the probe binds to the membrane, it signals the presence of a target complementary DNA sequence; meaning that the probe sequence exists in the original DNA. One of Southern's first uses of his blotting technique was to detect the presence or absence of the sickle cell mutation in human DNA samples.

In 1977, James Alwine, David Kemp, and George Stark at Stanford University in California developed the "northern blot," in which cellular RNA is used together with DNA probes to target a specific gene and determine whether that gene is expressed. This important advance allowed scientists to ascertain how the cell was using the information in its DNA.

Although northern and Southern blotting are valuable techniques, they employ only single DNA probes to examine DNA or RNA bound to a membrane. This limits analysis to one gene at a time. Scientists realized they could "flip" the process by binding multiple DNA probes directly in an array of fixed locations on a membrane, radioactively labeling DNA or RNA from a cell (so subsequent binding could be visualized), and then allowing the labeled nucleic acids to hybridize to the membrane-bound probes. This modified technique, which typically organized the probes gene by gene on a rectangular grid, allowed scientists to investigate hundreds of genes in a single experiment.

This was far beyond what had been done before, but it was still insufficient. In 1995, two teams of scientists—Robert Lipshutz, Stephen Fodor, and others at a company called Affymetrix, along with competitors Patrick Brown, Ronald Davis, and their colleagues at Stanford University—realized that the tens of thousands of genes in the human genome could only be analyzed if the array of probes was miniaturized. Their resulting DNA microarrays, or "gene chips," featuring DNA probes packed on a microscope slide, revolutionized biology much in the same way as miniaturization revolutionized the semiconductor industry (see Figure 5).

USING MICROARRAYS IN CANCER STUDY

Before the advent of microarray technology, an experiment would typically test a single-gene hypothesis: Is this gene involved in that particular disease? With DNA microarrays, scientists could shift from hypothesis-testing to "discovery science," in which research questions are much more open-ended: Which genes appear to be involved in that particular disease? At first, these two questions might not appear different from each other, but they are. By asking "which genes," scientists did not have to speculate

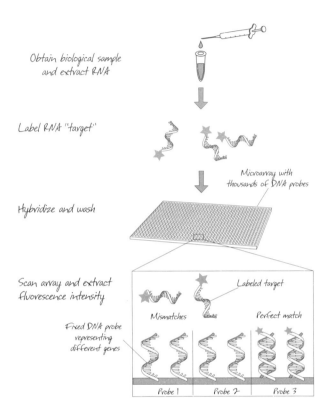

Obtain biological sample
and extract RNA

Label RNA "target"

Microarray with
thousands of DNA probes

Hybridize and wash

Scan array and extract
fluorescence intensity

Labeled target

Fixed DNA probe
representing
different genes

Mismatches

Perfect match

Probe 1   Probe 2   Probe 3

FIGURE 5: HOW DNA MICROARRAYS WORK

In DNA microarray expression analysis, an RNA "target" sample obtained from a biological sample is labeled with a fluorescent dye for detection, then applied to a DNA microarray gene chip on which DNA probes represent some or all of genes in the genome. The single-stranded RNA hybridizes with its complementary probe on the array to become a double-stranded nucleic acid. The degree of binding is assayed by measuring the fluorescence of each DNA probe to determine the level of expression of the corresponding gene.

about specific genes prior to conducting an experiment. Instead, the experiment itself could lead to unexpected discoveries about the genes involved in a particular disease.

One of the first applications of microarray technology was in the study of cancer. Between 1996 and 1998 many microarray studies focused on various aspects of the disease, such as comparing tumor tissue with the surrounding normal tissue to find differences in gene expression, or comparing the clinical stages of cancer to understand which genes distinguished them from one another. These early studies showed cancer to be a very complex disease that often exhibited confusing patterns of gene expression. In 1999, however, two studies, one of leukemia and the other of breast cancer, offered a new perspective on cancer— one based on genomics.

Leukemia is a less complex cancer of the blood-forming tissues typified by the overproduction of white blood cells in bone marrow and by the spread of abnormal cells to other organs. Todd Golub and his colleagues at the Dana-Farber Cancer Institute in Boston, Massachusetts, used DNA microarrays to examine two leukemia subtypes: acute lymphoblastic leukemia (ALL) and acute myeloid leukemia (AML). ALL is most common in children and characterized by an excess number of immature cells known as lymphoblasts, while AML is a cancer of the myeloid line of cells and typically affects adults. (Both lymphoblasts and myeloid cells are types of white blood cells.) Golub and his coworkers showed that genes expressed differently in ALL and AML (turned up in ALL and down in AML, or vice versa) and that their expression

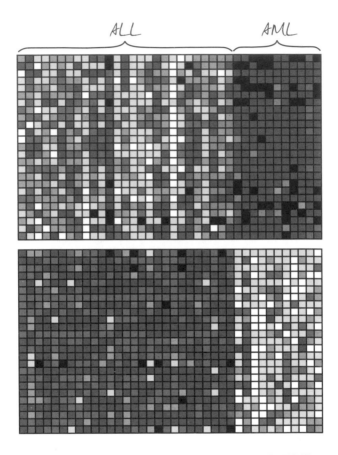

FIGURE 6: GENE EXPRESSION HEAT MAPS IN LEUKEMIA

Heat maps generated from DNA microarray data show gene expression values for a DNA probe in a given sample. Each row is a gene and each column is an individual patient. The largest values appear as "hot," or turned up, while the lower values are shown as "cool," or turned down. Intermediate values appear in varying shades. Here, two heat maps rendered in shades of black and white (two colors are used in the laboratory) show the expression levels for two leukemia subtypes: ALL and AML. The upper map shows genes turned up in ALL and down in AML; the lower map depicts the opposite.

patterns could be used as a *biomarker* to predict whether a patient had ALL or AML (see Figure 6).[4]

This experiment was transformative because it showed how genomic technologies could be used to create diagnostic tests. As microarrays became nearly indispensable in the scientific workplace, they provided an explosion of data and new approaches for studying nearly every known human cancer and a host of other diseases through gene expression. However, of the studies undertaken using this technology, breast cancer research is perhaps responsible for the greatest progress made to date.

ONE CANCER, MANY DISEASES

In a series of important papers published in 1999 and 2000, Charles Perou, Patrick Brown, David Botstein, and their colleagues at Stanford University described their use of microarrays to analyze breast cancer. What originally motivated them was the observation that women with seemingly identical breast cancer tumors exhibited different therapeutic responses and survival rates. Perou and his colleagues believed that this was due to differences in the tumors themselves that originated at the molecular level as variations in gene expression. They sought to determine whether breast cancer tumors could be separated into distinct subgroups based on their gene expression profiles and, if so, whether the subgroups correlated with clinically relevant outcomes. What they discovered would shed light on the nature of cancer.

Breast cells possess a series of hormone receptors on their

surfaces that respond specifically to estrogen and progesterone. When it comes to treating breast cancer, combining standard chemotherapy (which interferes with cell division) with therapies that block these hormones and their receptors can be more effective than chemotherapy alone. Using DNA microarrays, the team at Stanford discovered that similar-looking breast tumor samples could be separated into five subgroups based on their gene expression patterns, and that as few as 1,100 genes could define this separation. The primary differences among the groups were in the expression of genes associated with breast-cell hormone receptors. Also notable was that the associated patient groups had different prognoses and long-term survival rates; two of the groups had particularly poor prognoses.

In one poor-survival group, tumor cells expressed a gene called HER2 that encoded a cell-surface receptor similar to the estrogen and progesterone receptors. When the gene was turned on, it meant that the tumor cells had the HER2 receptor on their surfaces. As a result, the tumors could survive having estrogen and progesterone pathways blocked through standard therapies, since HER2 receptor activation could also stimulate breast cancer cell growth.

Coincidentally, scientists at the biotechnology company Genentech had been developing drugs that could block many types of cell-surface receptors, and among their collection was Herceptin, a drug that blocks the stimulating effects of the HER2 receptor. (In 2005, NIH-sponsored clinical trials showed that adding Herceptin to standard chemotherapy for HER2-positive tumors resulted in a 52 percent decreased recurrence risk for breast cancer.) Today, the standard of care for breast cancer includes testing for the HER2 receptor.

The patient group with the poorest long-term survival had the "basal-like" tumor subtype. Based on its molecular profile, this subtype is negative for estrogen, progesterone, and HER2 receptors, and tumors can grow in the absence of the hormone signals required by other tumor types. Consequently, the therapies that work with "receptor-positive" tumors fail in these basal-like tumors, resulting in the poor prognoses for those who have this type of breast cancer.

Aside from its specific application to breast cancer, Perou's work illustrates an important lesson that has implications for other cancers, and quite possibly for other diseases. What appears to be a single disease can often have hidden subtypes that can only be determined by examining the disease at the molecular level.

TOWARD PERSONALIZED CANCER CARE

The development of Herceptin as a treatment for HER2-positive breast cancers is just the first example of how genomics is ushering in new ways of treating diseases based on their molecular profiles. Along with new drugs are novel approaches that may help improve diagnostic and prognostic medicine and provide patients and physicians with valuable information for more effectively managing disease on a patient-by-patient basis.

Two diagnostic tools for breast cancer have come out of molecular profiling methods similar to Perou's and are becoming the standard of treatment for many patients. In 2002, Laura van 't Veer and her colleagues at the Netherlands Cancer Institute (NKI) in Holland, in partnership with the U.S.-based Rosetta Inpharmatics, Inc., used microarrays to profile gene expression in patients with early stage, node-negative breast cancer. They found a set of 70 genes that could be used to predict whether cancer would recur in patients whose breast cancer showed no evidence of metastasis to the local lymph nodes.

Based on their findings, which were later validated using a larger set of patients, scientists at NKI formed Agendia, a company that markets a 70-gene test called MammaPrint. At the time this book went to press, MammaPrint was undergoing a series of clinical trials in Europe and the United States. Data collected thus far suggests that the test will be a valuable addition to the arsenal of tools that oncologists are using to advise patients on their disease. In fact, the Food and Drug Administration (FDA) approved MammaPrint in 2007 for clinical use to determine the likelihood of breast cancer returning within five to 10 years after an initial diagnosis, making MammaPrint the first approved multigene genomic test to profile genetic activity.

Around the time MammaPrint was being developed, scientists at California-based Genomic Health were taking a different approach to the genomic study of breast cancer. After examining results of many published studies, they selected 250 candidate genes they believed were associated with breast cancer tumor behavior. The genes were analyzed in more than 400 patients from three independent clinical studies. When the smoke cleared, 21 genes strongly correlated with recurrence-free survival in women who had early stage tumors that tested positive for expression of the estrogen receptor, and for whom the cancer had not spread to the lymph nodes. (The shorthand here would be "ER+, node-negative breast cancer.") This gene set and the "recurrence score" calculated from it are now marketed as a diagnostic test called Oncotype DX.

Although both genomic tests search for different genes and employ slightly different ways of measuring expression levels, they reveal a fairly consistent picture of breast cancer. A 2006 study published in the prestigious *New England Journal of Medicine* showed that MammaPrint and Oncotype DX are better diagnostic indicators than the standard clinical criteria used to predict cancer outcome and survival. Rather than rely on tumor size, grade (how abnormal the cancer cells look), patient age, and other factors to direct therapy, these tests measure which genes are turned on—genes that can be linked to tumor metastasis. MammaPrint and Oncotype DX can also identify patients at the lowest risk for disease recurrence, allowing them to potentially avoid chemotherapy and its side effects.[5]

Thanks to genomics, breast cancer, once thought to be a more or less homogeneous disease, is now understood to be many diseases. Genomic tests have been developed to identify tumor subtypes, some of which now have specific and effective therapies. For other subtypes, further tests can make prognostic predictions. Scientists are also working on better therapies that target the unique features of each disease subtype.

Our increased molecular understanding of breast cancer has also aided in the development of directed therapies for other cancers. Gleevec, a drug that inhibits tyrosine kinase, an enzyme important in the activation of certain cancer cells, is used to treat chronic myeloid leukemia (CML) as well as a rare form of stomach cancer called gastrointestinal stromal tumor (GIST). The five-year survival rate for patients taking Gleevec is 95 percent.

In colorectal and lung cancer, there are now drugs that block the epidermal growth factor receptor (EGFR), which, like HER2 in breast cancer, promotes cell growth. Unfortunately, anti-EGFR drugs do not appear to work in cancers with a mutation in an oncogene called KRAS, found in high rates in colon, lung, and pancreatic cancers, as well as in leukemias. Candidates for anti-EGFR drugs are therefore routinely screened for mutations in KRAS before beginning treatment.

Although cancer study has been at the vanguard of the genomic revolution, nearly every other known disease and condition ranging from schizophrenia and autism to cardiovascular disease, diabetes, and infertility is being analyzed using genomic technologies, some of which will be discussed in the next chapter. Before we move on to them, however, we must look at other aspects of cancer that are changing how we look at nearly all diseases and their potential causes.

CANCER VIRUSES AND VACCINES

We have long known about cancer's genetic roots, and have for years studied environmental contributors to the disease, such as smoking and asbestos exposure. But there is another factor that plays a significant role in cancer: viral infection.

Viruses are an absolutely amazing form of life; that is, if they can be considered to be alive. Viruses do not share many characteristics with most living organisms. They have no cells, no mechanism to consume food or expend energy, and no way to reproduce on their own. They simply take over the cells of their hosts and cause them to carry on the processes necessary for the virus to exist and reproduce. Despite their inability to independently execute so-called living functions, viruses are the most common type of organism on earth.

When free of a host, a virus consists of little more than a nucleic acid—either DNA or RNA—"packaged" in a protective protein shell. It is basically inert, waiting for a chance encounter with a potential host, which could occur through an insect bite, a sneeze, a sexual act, or other form of transmission. Once it comes in contact with a host cell, the virus's shell acts as an elaborate molecular "syringe," attaching itself to specific proteins on the cell's surface, then injecting its nucleic acid (its genome), together with a few proteins, into the cell.[6] After the virus has hijacked the cell's machinery and created a safe haven for itself, it eventually directs the cell to make more viruses.

Viruses fundamentally alter how an infected cell operates, and this disruption can lead to diseases ranging from the common cold to the development of warts and even (eventually) cancer. People might find it surprising to learn that as many as 12 percent of human cancers can be attributed to viral infection. Viruses known to cause cancer include the human papillomavirus (HPV), associated with cervical cancer; hepatitis B and hepatitis C, linked to liver cancer; and Epstein-Barr virus (EBV), which is involved in three types of lymphoma. A growing number of viruses are being linked to cancer, and more and more cancers are being found to have a viral component. Although this information may seem a·bit terrifying, it actually offers hope for uncovering new ways of preventing cancer, most notably through vaccination.

One promising example of using a vaccine to dramatically reduce the incidence of cancer involves HPV. Of the nearly 100 subtypes of HPV that can infect people, approximately 60 of them are relatively harmless, causing only common warts on non-genital skin, such as on hands and feet. The other 40 are known as mucosal subtypes of HPV that affect the body's mucous membranes, which are the moist skinlike layers that line body cavities open to the outside, such as the vagina, anus, mouth, gastrointestinal tract, and lungs. Mucosal HPV subtypes are also called genital or anogenital because they typically affect the anal and genital areas. Some of the subtypes simply cause genital warts, but nine of them, the high-risk set, are associated with cancer of the cervix, anus, and other mucosal tissues.

Genital HPV is a very common virus, estimated to be as widespread as viruses that cause the common cold (not cold sores, but the yearly colds that are so prevalent). In the United States, more than six million people acquire an HPV infection every year, and between one-half and three-quarters of those who have had sex will get HPV in their lifetimes. Fortunately, nearly 90 percent of HPV infections are cleared by the body's immune system within two years and with little long-term effect. But infections that persist can cause problems. Cervical cancer is the fifth leading cause of death from cancer in women worldwide and the leading cause of cancer-related deaths among women in the majority of the developing world. More than 99 percent of cervical cancer cases are related to persistent HPV.

Scientists have discovered that other cancers can be caused by the same types of HPV that cause cervical cancer. This is true for half of the cancers of the vulva. Other genital cancers (cancers of the penis, vagina, and urethra) and some head and neck cancers (most often of the tongue and tonsils) may be related to the high-risk subtypes of HPV as well.

Evidence is also growing that HPV may be responsible for esophageal cancer, illustrating how environmental and behavioral factors might influence disease. Although rare, esophageal cancer is one of the fastest-growing cancers in the United States, with estimates of a nearly six-fold increase in prevalence since 1980. Since the same strains of HPV linked to cervical cancer have been implicated in esophageal cancer, one can speculate that changes in sexual behavior—perhaps reflecting societal (thus environmental) influences—might be associated with increased rates of cancers associated with the virus.

Genomic technologies are indispensable in exploring the connection between viruses and cancer. Specifically, DNA sequencing is providing information about viral genes that could lead to vaccines that might reduce the prevalence of virally associated cancers. One such success story is Gardasil, an FDA-approved vaccine developed by Merck & Co., and based on research of the gene sequence of HPV viral strains. Gardasil provides immunity to two high-risk HPV subtypes that, together, cause 70 percent of cervical cancer, as well as to two other subtypes responsible for 90 percent of genital warts. If widely used, this vaccine and others being developed have the potential to dramatically reduce the incidence of cervical cancer and cervical cancer deaths by an estimated two-thirds (and also reduce other HPV-linked cancers).

THE FUTURE OF CANCER THERAPY

Given that so many cancers arise spontaneously and that we continue to discover disease subtypes, there will almost certainly never be a "cure" for cancer. However, our knowledge of the disease has grown by leaps and bounds since the Human Genome Project began. Today, thanks to DNA microarrays, we can look at every gene in the human genome and know whether it is turned on or off in less time than it took in 1990 to analyze the expression of a single gene. We are also better able to link tumor expression patterns to the long-term survival or best therapeutic protocols for patients with a particular cancer. And the more we learn about the molecular and cellular mechanisms that drive the disease, the greater is our ability to arrive at new treatments to improve survival.

These and other dramatic improvements in research technology have inspired ambitious genomic-based projects around the world. One project currently under way is the Cancer Genome Atlas, a large-scale collaborative effort between the National Cancer Institute and the National Human Genome

Research Institute to create a comprehensive catalog, or atlas, of the genomic changes that occur in many cancers. The information will be placed into public databases for access by cancer researchers worldwide, furthering advancements in the clinical care of cancer patients and those at risk for developing the disease.

Although beneficial, using genomics to profile tumors and develop new treatments is a reactive approach to battling the disease. Scientists are also trying to be proactive in uncovering the spectrum of inherited mutations that increase cancer risk and the changes associated with tumor development and progression. Genomic technologies can be used to search for genetic alterations that run in families and thus help predict someone's likelihood to develop cancer. "Next-generation" DNA sequencing technologies allow scientists to read off the genome sequence of a tumor and compare it with the genome of a patient's normal cells to obtain a clearer picture of the molecular changes that occur as the tumor develops.

We have mostly focused our discussion on DNA- and RNA-based genomic techniques, but there are also important "spin-offs." For example, *proteomics* technologies are helping scientists examine the proteins encoded within the genome and search for differences between diseased and normal tissues. Although still a young science, proteomics has tremendous potential to influence how we detect and monitor the progression of disease. A great deal of effort in cancer proteomics, for example, focuses on identifying potential protein biomarkers in blood or urine that can be used for earlier cancer detection.[7] Proteomics also promises to shed light on the mechanisms that govern the development of cancer and its response to therapy.

Thanks to the data and tools made possible by the Human Genome Project in recent years, we are heading toward an exciting and promising future for cancer research, one in which we will be better equipped to detect disease, improve treatments based on our understanding of how disease alters a normal cell, and apply the most effective treatments to patients based on their cancer's unique molecular profile.

CHAPTER 4 NOTES

1. This is partly why chemotherapy drugs cause hair loss and gastrointestinal (GI) distress. Cells in these parts of the body, and elsewhere, are constantly dividing—in the hair follicles to keep hair growing and in the GI tract where the lining is replaced every few days.

2. Defining what constitutes a cure in cancer is very difficult, since another cancer can arise independently of a first manifestation of the disease. This is particularly common in some inherited cancers. The most common measure of being "cured" is a five-year, recurrence-free survival period following initial treatment.

3. The restriction enzymes come from bacteria, where they function as an immune system of sorts. Although bacteria never infect one another, they are susceptible to viruses. Restriction enzymes chop up viral DNA at specific sites to prevent the virus from taking over.

4. Although we most often refer to genes as being turned "on" or "off," a better description is turned "up" or "down" because the level at which a gene is expressed has potential consequences for a cell. If the mRNA level for a particular gene doubles—or is "up-regulated" two-fold—the resulting increase in the expression of the protein encoded by that gene can affect cell behavior. This suggests that gene expression is more analogous to a light controlled by a dimmer than by an on-off switch.

5. Not surprisingly, insurance companies are enthusiastic about MammaPrint and Oncotype DX because even at a cost of $4,000 each, the tests are far less expensive than a course of chemotherapy.

6. Viruses generally attack one type of cell and do not cross species, but there are exceptions. Influenza viruses can infect birds, pigs, and humans, and new influenza strains often develop when all three species live in close proximity to one another. Two or more viruses can infect the same cell, where their genetic material is mixed together, resulting in a new viral strain. The development of new viruses is also influenced by a rapid rate of mutation, which can even affect viruses that do not undergo genetic recombination. This is why developing specific vaccines can be so difficult.

7. Many single protein-based biomarkers assayed by blood tests, such as PSA in prostate cancer, CA125 in ovarian cancer, and CEA in colorectal cancer, are used in cancer therapy. However, each one on its own is a relatively poor cancer diagnostic because of the normal variation that exists in levels found in disease-free individuals. Nevertheless, once a diagnosis is made through other means, these markers help physicians monitor the disease's progression. Proteomics-based technologies might one day lead to more sensitive protein-based biomarkers that could provide better early detection.

CHAPTER 5

Genes and the Roots of Disease

Although genomics has thus far had its greatest impact on the study of cancer, it has transformed all biomedical research, opening up new avenues of study, even in those diseases for which we have already found a genetic "smoking gun." Today we are better equipped than ever to discover and interpret the root causes of diseases, but even with all that the tools of modern biology and genomics can offer, significant challenges remain.

The principles of Mendelian genetics have been essential in helping scientists understand the link between genes and disease. Knowing how genes are inherited, whether they are dominant or recessive, and how mutated genes can cause cells to go awry, among other things, provides valid and meaningful explanations to the origin of certain diseases. However, geneticists recognized early on that tracking traits and gene expression through inheritance was not quite so simple. Yes, some traits could occur in families with a greater likelihood than expected to occur by chance, but in many cases it was not with the precise frequency expected under the principles of Mendelian genetics. Traits could sometimes even skip generations. To account for this, geneticists invented the idea of "reduced penetrance," which suggests that even dominant genes do not always manifest themselves as measurable traits, implying that other factors are at play.

Today, a much more complicated picture has emerged of genetics and the link between one's *genotype* (or genetic makeup) and *phenotype* (one's observable characteristics). Genes interact with one another, are influenced by alterations to DNA and the proteins that provide structure to chromosomes, and interact with the environment in creating or mitigating the development and severity of genetically linked diseases. Genes are still a driving factor in disease development, but they are only part of the puzzle. In this chapter, we will look at a number of diseases—some that follow Mendel's rules of inheritance and other more complex, multifactorial diseases. To do that, we must first look more closely at how traits are passed from parents to offspring and how modern genetics lets us track down individual genes through inheritance.

BEYOND MENDEL: MODERN GENETICS

One of the early goals of the Human Genome Project was to find landmarks, or unique genetic markers, across each of the 46 human chromosomes. In addition to helping scientists correctly reassemble the genome sequence, these markers are indispensable for seeking out genetic causes of disease. With the genome sequence finished, we now have millions of precisely mapped DNA segments that vary slightly among individuals, often by only a single base. These minute variations in sequence are immensely useful when examining patterns of heredity. For example, by comparing the genetic markers you carry on your chromosomes with those of your parents, it is possible to determine which of your chromosomes came from which parent. Even more interesting is that you can also uncover the portions of your chromosomes that came from your grandparents, your great-grandparents, and so on.

This deduction is possible thanks to the way gametes are produced in sexually reproducing organisms. If you recall from chapter 1, Walter Sutton observed that in the final stages of sperm and egg formation, chromosomes in the original cell split but did not duplicate. What Sutton had witnessed was a special type of cell division called *meiosis*. Ultimately, this process results in the creation of not two *diploid* cells (with 46 chromosomes each), as occurs in mitosis, but of four *haploid* cells, each with only one copy of the body's 23 chromosomes. During successful fertilization, when sperm and egg cells fuse, the resulting cell, or the *zygote*, will contain a diploid genome with the proper number of chromosome pairs.

Just before the chromosomes split during meiosis, a critical stage called *recombination* takes place, during which chromosomes come together and "cross over," swapping pieces of DNA between them (see Figure 7). The resulting chromosomes are a patchwork composed of pieces of the chromosomes of the parent cell. This is why every sperm and egg cell that the body creates is unique and why genetic variation exists in all of us.

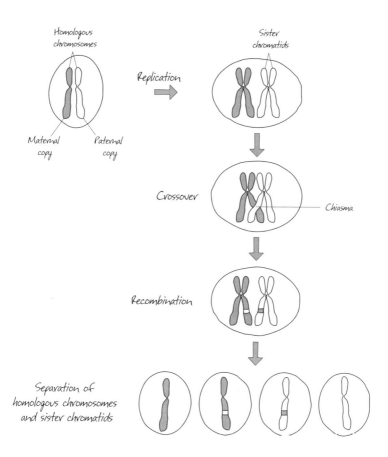

FIGURE 7: CHROMOSOMAL RECOMBINATION

Each diploid human cell contains two copies of each chromosome, a maternal and a paternal copy. During meiosis, the process of making sperm or egg cells (gametes), these copies pair with each other and replicate to create four copies. Following replication the paired chromosomes recombine, exchanging segments between sister chromatids. The (mosaic) chromosomes are then assorted into haploid gametes. This simplified example shows a single crossover between two sister chromatids.

To better illustrate chromosomal recombination, let us use the analogy of a simple child's toy. Imagine that each copy of your chromosome 1 pair is built out of LEGO blocks. The copy you inherited from your mother is made up of red blocks, while the copy from your father comprises blue blocks. During meiosis, these red and blue "chromosomes" break apart in large chunks that are swapped across each other so that in the end, the sperm (or egg) cell that your body has now produced contains only one copy of chromosome 1. However, that copy is a mosaic of red and blue—a mix of your mother's and your father's DNA.

Meiotic recombination occurs in every generation. Your parents' chromosomes were mosaics of their parents' chromosomes and so on. Genetically, each of us is a patchwork of the chromosomes of our ancestors. Scientists use meiotic recombination and genetic markers to trace these "patches" of inheritance back across generations, looking for genetic variations in relatives who had or have a particular disease. Finding such variations narrows the search for causative genes to small regions of the genome and thus dramatically speeds up the discovery process. This, in essence, is the basis of genetics, and the three examples that follow show how genetics and genomics are used together to better understand disease.

HUNTINGTON'S DISEASE: AN AUTOSOMAL DOMINANT DISEASE

Huntington's disease, also known as Huntington's chorea or HD, is a neurodegenerative disorder first described in 1872 by

American physician George Huntington. HD is characterized by chorea (from the Greek *choriea* for "dance"), which describes a pattern of repetitive, abnormal movements. The disease, affecting approximately seven out of every 100,000 individuals of Western European descent, can be found in concentrated numbers in geographically localized regions.[1]

A devastating illness by all accounts, HD is nevertheless important to discuss from a research standpoint because it was one of the first diseases in which a cause was identified using resources developed during the mapping and sequencing of the human genome. In many ways, the identification of the HD gene was a case of having the right genomic tools available at a time when people were ready to use them.

How HD affects individuals varies, but it features a characteristic progression of symptoms. Earliest signs include a general physical restlessness, small unintentional and often uncompleted body motions, a lack of coordination, and an unsteady gait. These physical symptoms worsen, with rigidity, repetitive motions, and abnormal posturing becoming dominant. Eventually, any action requiring muscle control is affected, resulting in instability, abnormal facial expressions, and difficulty in chewing, swallowing, and speaking. Along with the physical manifestations comes a progressive impairment of cognitive abilities, with behavioral and psychiatric problems ultimately leading to dementia. Although HD itself is not fatal, full-time patient care becomes necessary in the disease's later stages and complications reduce life expectancy to about 20 years after the onset of symptoms.

HD runs strongly in families and is inherited dominantly, so the child of an affected person has a 50 percent risk of inheriting the disease. Physical symptoms can begin anytime, although the average age of onset is between 35 and 44 years. Because HD often strikes later in life, multiple generations of a family can be afflicted with the disease.

One of the most valuable assets a scientist can have when searching for disease-related genes through heredity is access to a very large family in which a given disease occurs and for which good family records exist. Fortunately, during the early study of Huntington's disease just such a family was found—indeed, the world's largest—at the right time and by the right person.

It is almost impossible to talk about HD research without mentioning the work of neuropsychologist Nancy Wexler. When it came to understanding the disease, Wexler had a personal motivation to research the genetic basis for the disorder. She had watched her uncles and mother succumb to the disease. Her father, Milton Wexler, a Los Angeles psychoanalyst, was the founder of the Hereditary Disease Foundation. To raise awareness of HD, the foundation had traveled to the village of San Luis in the Lake Maracaibo region of Venezuela to film people who were suffering from the disease at a surprisingly high rate—the highest in the world. After seeing the film, Wexler obtained research funding from the National Institutes of Health in 1978 and headed for San Luis to hunt down the HD gene.

Although the villagers were initially hesitant to cooperate with outsiders, Wexler's personal link to the disease opened doors and allowed her and her colleagues to convince people to help sort out family connections and donate blood samples. Wexler eventually compiled a pedigree—a family tree showing the presence or absence of a trait; in this case, HD—of nearly 10,000 people, collecting more than 2,000 blood samples in the process.[2]

In 1983, Harvard molecular biologist James Gusella, Indiana University's Michael Conneally, and the Massachusetts Institute of Technology's David Housman agreed to analyze Wexler's DNA samples. Using newly developed genetic markers, the team looked for those variants that traveled with the HD gene and appeared in villagers affected by the disease. It was a trial-and-error effort that was expected to take years as hundreds of markers were screened against all the collected DNA samples. But to everyone's surprise, the team struck gold with the third marker they tried. Located on chromosome 4, it differed between those who had HD and their healthy relatives.

Although finding the marker was an incredible breakthrough, it would take another 10 years to find the gene called Huntingtin that, when mutated, causes HD. The Huntingtin gene is *autosomal dominant*, meaning that it is a dominant gene that does not reside on either the X or Y chromosomes. (Remember, we have 22 pairs of autosomes and two sex chromosomes.) It was also the first autosomal dominant gene found, and its discovery led to the development of genetic tests for HD that can be performed before the onset of symptoms, at any age, even before birth. Although invaluable, these tests have understandably raised ethical issues involving reproductive choice.

Surprisingly, the mechanism linking the HD gene and the

disease itself is not yet fully understood, but a number of factors are known. A mutation in the Huntingtin gene causes the production of an abnormal form of the protein huntingtin that, over time, produces cellular and anatomical changes in the brain, which eventually lead to the development and progression of HD. Although there is no cure for HD, there are now treatments to at least relieve some of its symptoms.

The story of Huntington's disease reveals important lessons that can be applied to other genetic diseases. The first is that the most important resource available to scientists for mapping the location of disease genes are the affected individuals and their families. A second lesson is that finding a disease-causing gene often does not lead to a cure or treatment. Instead, it is when the real work begins. To learn what the diseased version of the gene does and to then compensate for it or correct the damage it does can take longer than finding the disease gene itself. Again, a treatment might not be found.

Last, and perhaps most important, finding a disease-causing gene can raise questions that go beyond science and medicine. A challenge to having a presymptomatic test available for the HD mutation is that it also calls for genetic counseling by those trained to help people understand the test, as well as the benefits and risks of knowing their potential for developing a disease that has no effective treatment. Counselors can also help people comprehend the risks for their children or future children and make informed decisions about reproduction.

Looking back at what it took to find the gene for HD before the HGP map was available shows us how far we have come in the hunt for disease-causing genes. Today, with the entire genome at our disposal and technologies that can test as many as one million genetic markers at a time, mapping HD and finding the gene would have happened much more quickly. But the effort would still have required gathering together the necessary families and biological samples; and after finding the marker, painstakingly working to discover the relevant gene and the role that its mutation plays in HD. Clearly, aspects of biomedical research have remained largely unchanged despite the genomic revolution. Simply put, the hunt for genes will always require hands-on detective work.

CYSTIC FIBROSIS: AN AUTOSOMAL RECESSIVE DISEASE

Huntington's disease is an example of a dominant trait; if you inherit one copy of the mutated gene from either parent, you will develop the disease. But most diseases are not dominant, and this next example is an *autosomal recessive* disease and, as such, requires a person to inherit mutated gene copies from both parents to manifest the disease.

One of the most common types of chronic lung disease in children and young adults is cystic fibrosis, or CF, a deadly illness that causes the buildup of thick, sticky mucus in the lungs and digestive tract. CF generally appears in infancy or early childhood, and its symptoms include salty-tasting skin, poor growth and low weight gain, excess mucus production, coughing, and shortness of breath. The inflammation associated with mucous buildup can also promote the growth of harmful bacteria, leading to pneumonia as well as to a wide range of other health problems. Despite available approaches for managing the disease, average life expectancy for someone with CF is only about 35 years.

Most prevalent in people of European descent, cystic fibrosis has a strong genetic component and manifests when a person receives two copies of the associated mutated gene, CFTR. In the United States, it is the most common deadly inherited disorder affecting Caucasians—one in 3,300 Caucasian children are born with the disease compared with one in 15,000 African American children and one in 32,000 Asian American children.

Finding the CFTR gene was quite a challenge, in part because a recessive gene is more difficult to locate than a dominant gene. In the 1980s, a series of genetic studies of families carrying the disease first linked CF to chromosome 4, but then later to chromosome 7, where the evidence was greater. However, scientists were unable to precisely locate the associated gene by tracking inheritance patterns in families, which had proven successful for the Huntingtin gene.

Eventually, they developed a new approach for identifying disease genes, one called *positional cloning*, or "reverse genetics." Instead of starting off with a trait (or disease), scientists would first uncover genes in a given chromosomal region, ascertain what they did, then associate them and their functions back to a trait.

The discovery of the gene for CF was announced in July 1989 by scientists at the Hospital for Sick Children in Toronto, Canada, led by Lap-Chee Tsui and John Riordan, along with Francis Collins at the University of Michigan. Their approach involved first identifying a segment of DNA on the region of chromosome 7 to which

CF had been mapped and using as a probe to search a "library" of what are known as *cDNAs*. (Scientists extract mRNAs from cells and convert them to cDNAs, which are complementary to their original mRNAs.) The cDNAs were then sequenced and used to determine what proteins the mRNAs encoded.

One of the proteins uncovered was a chloride ion channel that positions itself in a cell's membrane and regulates the flow of calcium into and out of the cell. This protein is important in creating sweat, digestive juices, and mucus. It turned out that the CFTR gene encodes this protein, and its mutation results in the symptoms of cystic fibrosis.

As with HD, finding the CFTR gene did not lead to a cure, although significant medical advances have been made. Genetic tests can identify carriers of the CFTR mutation, paving the way for genetic counseling and informed reproductive choices, and new drug therapies are extending the lives of individuals suffering from the disease.

DIABETES: A POLYGENIC, MULTIFACTORIAL DISEASE

Applying the principles of genetics is a powerful way to find genes associated with diseases like Huntington's disease or cystic fibrosis, where single gene mutations are the primary drivers causing the disease. However, as scientists expanded their hunt for genes, they discovered that many diseases, and numerous physical traits such as eye and hair color, are not caused by a single gene, but by many genes acting together, with environmental factors often contributing significantly. Clearly, the inherent complexity of these diseases, called *polygenic disorders*, precludes a simple answer to causality.

Diabetes mellitus, more commonly referred to as diabetes, is an excellent example of a polygenic disorder. Diabetes involves problems with the manufacture and regulation of a hormone called insulin, which is produced in the pancreas and used by the body to convert glucose, a type of sugar, into energy. The disease occurs when the body either does not make enough insulin or is unable to use insulin efficiently, or both. As a result, glucose accumulates in the bloodstream, which can cause cardiovascular disease, kidney failure, retinal damage (often leading to blindness), poor wound healing (particularly in the feet where gangrene can occur), and microvascular disease, which can cause erectile dysfunction.

In the developed world, diabetes is the leading cause of adult blindness and non-traumatic amputation. Diabetes-related kidney failure is the main illness requiring dialysis in the United States, where an estimated nearly 8 percent of the population suffers from the disease. Almost one quarter of these people are unaware that they even have diabetes.

Diabetes is actually a family of diseases with two primary types: type 1 and type 2. Type 1 diabetes results from the body's failure to produce insulin. An estimated 5 percent to 10 percent of Americans who are diagnosed with diabetes have type 1, better known as juvenile diabetes or insulin-dependent diabetes mellitus (IDDM) because it generally begins in children whose bodies cannot produce insulin and who must therefore take insulin injections to control the disease and its symptoms.

Type 2 diabetes, known as adult-onset diabetes, obesity-related diabetes, or non-insulin-dependent diabetes mellitus (NIDDM) arises when, in combination with a relative insulin deficiency, the body fails to properly use insulin. The deficiency often progresses during the course of the disease, eventually requiring sufferers to rely on insulin injections to control the disease. A third subtype, gestational diabetes, can occur in pregnant women, but it generally resolves following birth.

Most Americans diagnosed with diabetes have type 2; in the United Kingdom, 85 percent of diabetics have this type. These people have often spent many years in a prediabetic state, with blood glucose levels higher than normal, but not high enough to classify them as diabetics. In the United States, an estimated more than 57 million people are prediabetic, and most of them will eventually develop type 2 diabetes.

Diabetes is believed to be at least partially inherited; however compelling evidence indicates that environmental factors influence the disease's development and progression. Type 1 diabetes appears to be triggered by a viral or other infection or, less commonly, by stress or environmental exposure to certain chemicals or drugs. Some of the genetic susceptibility to the disease has been traced back to the *human leukocyte antigen system (HLA)*, which involves proteins expressed on the cell surface and is used by the body to distinguish between its own cells and foreign cells. This suggests that type 1 diabetes is in large part an autoimmune disorder. But even people with the most susceptible HLA genotype still require an environmental trigger.

There is a much stronger pattern of inheritance for type 2

diabetes; nearly 25 percent of those with the disease have a family history of it. Having a parent or sibling with type 2 diabetes greatly increases the chance you will develop the disease, and the risk grows as the number of diabetics in your family increases.

Despite extensive genetic studies, including the discovery of more than 10 genes associated with an increased risk of type 2 diabetes, no single gene or even combination of genes has been identified for type 2 diabetes. The fact is, none of the mutations discovered so far have nearly as much predictive power for diabetes as does a high body mass index (BMI), a measure of obesity. Unfortunately, given the obesity epidemic in the United States, where the number of prediabetic and type 2 diabetic individuals is projected to grow dramatically, this aspect underlying the disease poses a serious health concern for individuals and society at large, particularly since research has not yet led to better treatments or genetic screening tests for diabetes.

Given its polygenic and multifactorial nature, diabetes is the prototypical example of the vast majority of human diseases, which makes research all the more challenging. Other such diseases include asthma, autism, cardiovascular disease, schizophrenia, inflammatory bowel disease, and autoimmune diseases like multiple sclerosis. In fact, complex diseases greatly outnumber the "simple" Mendelian diseases, implying that although genetics and genomics have yielded tremendous medical insights, the greatest challenges lie ahead. And as we will see, there are factors beyond genetics that can further muddy the investigative waters.

GENES, ENVIRONMENT, OR BOTH?

Scientists have come a long way from Mendel's peas and Morgan's mutant fruit flies. We now know that regardless of whether polygenic or monogenic, nearly all human diseases involve a complex interaction between genetic and environmental factors. This interaction is further complicated by the sheer number of environmental elements that can influence human health—tobacco smoke, drinking water, asbestos, viruses, radon, mental abuse, radiation, toxins, chemicals, and so on—as well as the fact that a person's genetic makeup can alter the response to environmental factors.

Some diseases, such as Huntington's disease, are predominantly influenced by genetics, with environmental factors playing only a minor role by modifying the severity or possibly delaying or accelerating the course of the disease. Other diseases, such as diabetes, are strongly influenced by environmental factors, with genetics playing a role in determining one's susceptibility to the disease, but unable, as far as we know, to cause the onset of a disease except in possibly the rarest cases. For example, the severity of cystic fibrosis is affected by several environmental factors, including exposure to tobacco smoke, poor nutrition, and stress. Cardiovascular disease and high blood pressure are greatly influenced by one's diet and general weight.

Then there are those environmental factors known to play an important role in disease. Nearly every cancer has a recognized environmental contributor. A risk factor for skin cancer is excess exposure to sunlight; for colorectal cancer it is a high-fat, low-fiber diet. Lung cancer, which kills more Americans every year than any other cancer, is heavily influenced by exposure to tobacco smoke, making it one of the most preventable cancers. If everyone quit smoking, the rates of death from lung cancer (and other cancers) would fall dramatically. Of course, this does not mean that lung cancer would go away. Genetic factors contribute to lung cancer, and other environmental factors increase lung cancer risk, including diesel exhaust, silica dust, radon gas, and the use of talcum powder. This type of interplay between genetic predisposition and environment contributes to nearly all cancers.

The bottom line is that many elements in our environment, from naturally occurring compounds like radon to man-made chemicals such as dioxin and PCBs, can influence our health and the development and progression of disease. And, in many ways, our understanding of these effects is still in its early stages. Toxicology is largely an empirical science, whereby toxicity for a compound is estimated by exposing animals to increasing doses and measuring deleterious effects. The incorporation of genomics into toxicology research has led to a new field known as *toxicogenomics*, which involves studying how cells and tissues react at the molecular level to toxic substances. The hope is that we will develop more sensitive and accurate tools to study environmental contributions to disease.

GENETICS AND INFECTIOUS DISEASE

In the study of disease, we long ago learned that not everything is as it appears. Just when you think you have finally figured out something, surprises surface. Some diseases that we once thought of as being caused solely by environmental factors, such as bacterial or viral infections, are in truth influenced by genetics. Genetics can even influence something as simple as whether a cut becomes infected. If, for example, two people with open wounds were exposed to the same bacteria, the relative chances of either one (or neither or both) developing an infection would depend in part on their genetic background and the antibodies they are carrying from prior infections.

A more serious example of genetics influencing environmental factors is the human immunodeficiency virus (HIV), a tremendous health concern and the subject of intense study since its identification as the cause of acquired immune deficiency syndrome (AIDS). AIDS is a disease that progressively reduces the effectiveness of the human immune system, leaving individuals susceptible to opportunistic infections and tumors; and it is these secondary infections that eventually kill people with the disease. In 2007, the worldwide AIDS pandemic killed an estimated 2.1 million people, including 330,000 children; another 33.2 million people contracted the disease. Efforts have been under way since the early 1980s to develop a preventive HIV vaccine, but to no avail. Nevertheless, there have been some interesting findings, including the 1996 discovery by a group of scientists in Belgium of a mutation in a gene called CCR5, which appears to protect carriers of that gene against HIV infection.

By analyzing the genomic sequences of a large number of people, scientists discovered that the CCR5 mutation is a relatively recent one (about 700 years old) and occurs in only about

10 percent of people, those being almost exclusively of European descent and mostly from far-northern Europe. Given the historical events surrounding the timing of the mutation's appearance, scientists first speculated that the CCR5 mutation might have arisen about the time of the bubonic plague, and that carriers of the mutation might have had some advantage in resisting the disease. But subsequent experiments have shown that the CCR5 mutation that protects against HIV infection does not protect against the plague. Surprisingly, carriers of the mutation are more likely to be infected by the West Nile virus.

This example illustrates how one's genetic background can set the stage for susceptibility to environmental factors that can, in turn, contribute to the development of various diseases. Additionally, environmental factors can cause mutations beyond those that are inherited, thus further increase susceptibility. Or, elements in the environment can stress a person's cells in such a way that small inefficiencies or shortcomings in the cells' systems are pushed further and further from their optimal functioning. Eventually, the systems break down and those failures are what manifest as disease.

IT TAKES A VILLAGE: SOCIAL ENVIRONMENTS AND DISEASE

When we study complex diseases, our efforts to separate environmental factors from genetic factors are complicated by issues such as ethnicity, income level, socioeconomic status, and other considerations that can potentially confound the search for a "cause." Although family studies are important for investigating diseases with a strong genetic basis like HD, they are not always ideal for complex diseases. The better research approach for these is known as the *cohort study*, a type of longitudinal study designed to collect health data from a defined group of people over a period of many years.

One of the best-known cohort studies is the Framingham Heart Study (FHS), which began in 1948 with the goal of identifying the common factors that contribute to cardiovascular disease. The study involved an original cohort group of 5,209 Massachusetts men and women, 30 to 62 years old, who had no known symptoms of heart disease nor had suffered a heart attack or stroke. Participants agreed to complete questionnaires and undergo physical exams every two years for the entire duration of their participation. In 1971, a second generation of subjects was enrolled, consisting of 5,124 of the original participants' adult children and their spouses. And in 2002, a third generation was enrolled, eventually numbering 4,095 grandchildren and their spouses.

A great deal of what is now common knowledge regarding heart disease, such as the effects of diet and exercise, the role that obesity plays, and the protective effects of low doses of aspirin, was discovered through FHS data analysis. But one of the study's most intriguing findings was made through secondary data analysis by Nicholas A. Christakis of the Harvard Medical School and James H. Fowler of the University of California, San Diego, who wanted to explore whether obesity was "contagious" and spread through social networks. They used FHS data to construct a network that linked people together based on whether or not they were friends. This was possible because so many participants had been asked to name their friends to facilitate follow-up during the study.

When looking at pairs of mutual friends, Christakis and Fowler found that if one friend became obese during a given time interval, the other friend's chances of following suit increased by 171 percent. Among pairs of adult siblings, if one sibling became obese, the chance that the other sibling would become obese increased by 40 percent. Apparently, friendship had a much greater effect on one's likelihood of becoming obese than did family ties or family genetics. In fact, obesity was found to cluster in communities. The chance that a friend of a friend of an obese person would also be obese was about 20 percent higher than one would expect by chance, implying that obesity has a very strong social component and that this aspect of one's "environment," along with other environmental factors, cannot be ignored when analyzing the causes and treatments for the disease.[3,4]

Not surprisingly, scientists have also discovered genetic contributions to obesity. While conducting genetic analysis on a large cohort of individuals, a consortium of scientists in the United Kingdom discovered that a common variant of a gene called FTO, found on human chromosome 16, was associated with an increased risk for obesity. But the genetic risk for those who carried the gene variant was still far smaller than the risk that came from having an obese friend.

Although many lessons can be drawn from Christakis and Fowler's work, the most important one is that most diseases are more complex than we imagined—driven by a number of envi-ronmental, societal, genetic, and even psychological risk factors, among others. Ironically, obesity itself is a risk factor for diseases ranging from heart disease and diabetes to cancers of the colon, breast (in postmenopausal women), endometrium (the lining of the uterus), kidney, and esophagus, to mention a few, meaning that each of these diseases has a social component, too.

PERSONALIZED MEDICINE

Despite any research limitations that this chapter discussion suggests, the science of genomics is invaluable and its tools indispensable in tackling complex diseases. Although many causes and cures have yet to be found, our disease diagnostics, treatments, and management have been profoundly redefined, and the examples are as unique as they are numerous.

One example is AlloMap, a gene expression test developed by a company called XDx (Expression Diagnostics) to identify heart transplant patients at risk for organ rejection. AlloMap examines a patient's white blood cells using technologies similar to microarray tests described in chapter 4. Normally, these immune cells help defend the body against infection by identifying and attacking foreign entities, which can include organ transplants. Doctors typically perform risky invasive heart biopsies to determine whether white blood cells are attacking transplant tissue. AlloMap is a simple blood test that detects changes in the cells' gene expression levels. In this way, patients at risk for organ rejection can be identified and started on therapies sooner, thereby increasing their chances for surviving the transplant. Approved by the Food and Drug Administration in 2008, AlloMap has become part of the standard of care for heart transplant patients in some hospitals.

SYSTEMS BIOLOGY: MORE THAN THE SUM OF OUR PARTS

Until recently, modern biology was primarily driven by the Central Dogma of Molecular Biology. Genes are transcribed into RNAs, which are then translated into proteins, each of which is associated with a particular function in the cell and a unique, observable phenotype. This is why the search for mutated genes has dominated much of human biology; finding "the" gene meant finding the cause of disease.

Today, biological studies are being increasingly dominated by a new approach known as *systems biology*, which recognizes that biological systems are governed by intricate networks and pathways than by single biological components, such as genes or proteins. For some diseases, the failure of a single component (for example, the mutation of a gene) can be enough to cause the entire system to break down. For others, a collection of genetic and environmental factors contributes to the development of a disease state.

A useful analogy for illustrating systems biology is your car and the systems that keep it running smoothly (the electrical system, the cooling system, the engine system, etc.). An old battery, a few cracked sparkplug wires, or a partially clogged fuel filter might not in and of themselves prevent your car from running. But taken together, these parts could easily bring things to a halt. Diagnosing the cause of a systemic automotive problem can challenge even the best mechanic. In terms of biology, sorting out the systemic causes of human disease represents one of the greatest scientific challenges of the twenty-first century.

Systems biology is also exciting because it brings together many scientific disciplines, including genomics, medicine, mathematics, and computer science. It borrows some techniques and invents others to give us a way to take the genes in the "parts list" we have found in the human genome sequence and, eventually, assemble them into a "wiring diagram" that we can use to better understand cells and their functions.

CHAPTER 5 NOTES

1. Rates of Huntington's disease in people of African and Asian descent are only about one in 100,000.
2. The study would ultimately include 18,000 people and more than 4,000 blood samples.
3. There was no effect on a person's weight when a neighbor gained or lost weight and, as noted, family members had less influence than friends. The effect was greatest between mutual close friends; however, even friends separated by hundreds of miles were affected. Why this was so is not clear. Christakis speculated that friends might influence each others' perception of fatness, making it more acceptable to be obese if one's close friend is obese. Christakis and Fowler suggested that this social factor might be partly responsible for rising obesity rates in recent years.
4. In the same analysis of the Framingham Heart Study data, Christakis and Fowler found that smoking behavior and happiness also exhibited clustering in social networks.

Chapter 6

The Genomics of Evolution

Humanity has wrestled with the question of its origin from its earliest days, and nearly every society and religious tradition has a creation story describing how human beings came to assume their place in the panoply of creatures with which they share the earth. Charles Darwin's *The Origin of Species* inspired us to consider this question scientifically, by exploring the biological roots of our existence.

As far back as Darwin's time, scientists had been examining the physical characteristics of living organisms and fossil evidence, concluding that our closest relatives in the animal kingdom were the great apes, including gorillas and chimpanzees, and that we separated from these creatures millions of years ago, evolving into a distinct species. Paleontologists in Africa and elsewhere searched for fossils that could explain how this divergence occurred. They uncovered striking evidence showing that the development of modern humanity was a gradual evolution over millions of years and with humble beginnings in East Africa. Today, the Human Genome Project and its related technologies have provided us with a different kind of "fossil record"—an evolutionary history as captured in our genomes.

EVOLUTION'S NOTEBOOK

In chapter 1, we discussed how Charles Darwin's comparison of the similarities and differences in the Galápagos finches helped lead him to his theory of evolution by natural selection. To model this evolution, Darwin built what amounted to a family (or *phylogenetic*) tree for the different bird species, representing his hypothesis of how the finches related to one another. The tree began with a single parent species that migrated from the mainland and showed relationships to groups of finches that spread to populate the islands. As the finches spread, each island presented its own unique challenges to survival. In response, these groups evolved into new species through "descent with modification," adapting to local conditions.

In addition to outlining relationships among the species, Darwin's analysis also provided a basis for understanding how finches spread across the Galapágos, since the physical characteristics represented in the family were was closely tied to the geography.

At the heart of any phylogenetic tree, and of phylogeny itself, is a simple but powerful idea, *parsimony*—the assumption that nature makes as few changes as possible as new species evolve from old ones. This concept of economy complements Darwin's theory, which holds that change in a species (now thought of as evolutionary change) is a gradual, stepwise process. It also puts forward what we now know to be true: Nature reuses what works.

Consider land mammals. Most walk on all fours, or at least something approaching that. But the great apes are also able to amble about on their hind legs, suggesting, based on this one observation alone, they are likely our closest relatives. However, what allow humans to stand fully erect on our "hind" legs are the differences present in the structure of our limb bones and hips compared with those of animals that walk on all fours.

These bone structure differences, along with the progression in development of human hips and limbs, are something we see in mammalian skeletons, with the apes collectively falling between humans and four-legged animals. These physical modifications are recorded as changes in genes that govern limb formation. In other words, our genomes contain a record of our evolution as a species. Here, too, parsimony is a guide in tracing our lineage, since it makes sense to assume that the species whose genome sequences are most similar to each other are the species that are most closely related.

The physical characteristics of past and present species have allowed scientists to develop remarkably accurate phylogenetic trees that trace the development of life on earth, making today an exciting time for evolutionary biologists. Genomic tools, techniques, and data are opening up new areas, and reopening old areas, of exploration into who we are and where we came from, how similar we are to each other, and how we as a species fit within the continuum of life.

OUT OF AFRICA: GENOMIC DIVERSITY

Earth's fossil record tells an astonishing story of human evolution, involving multiple waves of prehuman species arising in Africa and spreading throughout various regions in Europe, the Middle East, and Asia. But as compelling as it was, fossil evidence was subject to different interpretations that often led to disagreements about where modern humans first arose. Fossils found outside of Africa led some scientists to speculate that humans may have arisen in multiple locations.

However, the consensus that eventually emerged was that between 100,000 and 200,000 years ago *Homo sapiens* evolved in Africa to become what we consider anatomically modern humans, and that a relatively small group left the continent about 60,000 years ago, arriving at destinations where they ultimately pushed out many earlier prehuman populations (including Neanderthal and *Homo erectus*) who had made the same journey throughout the previous millennia.

As genomic science has grown, so has genetic evidence that supports an African origin for modern humans. The most compelling data comes from the search for gene variation within the human genome. Although any two humans are nearly 99.9 percent similar in their genomes, that one tenth of 1 percent difference is not inconsequential given that our genomes contain more than three billion bases. This small difference is equivalent to an average of one base change for every 1,000 bases, or more than three million small changes. We have discussed how scientists rely on these differences to map human disease; but variations can also be used to map our evolution and early migration history.

In 2002, the International HapMap Project set out to conduct a survey of variation in the human genome by looking for single base changes, or *single nucleotide polymorphisms (SNPs)*, and the precise locations where they occur along the genome sequence. The goal was to produce a *haplotype* map of common variants (those appearing in at least 1 percent of the populations surveyed) that could then be traced through families in a search for genes linked to disease.

One of the side benefits of assembling this map was the extensive data gathered on millions of variant sites in the genome and their distribution among populations. Project scientists found many more polymorphisms in African populations than in populations from other regions. Equally fascinating was that plotting these patterns of genetic variation on a world map essentially revealed the history of human migration: Humans had emerged from Africa, traveled into the Middle East, then spread into Europe, across Asia and on to Australia; and, later, from Asia to the Americas. The evidence is rooted in the fact that the older a populations is, the greater is its genetic variation.

Imagine an initial population arising in East Africa 150,000 years ago and expanding to cover large portions of the continent. As the population grew, the natural processes of mutation introduced genetic variation. Consequently, each small group in different geographical areas developed genetic variations that distinguished them from other groups. Over a span of 50,000 or so years, the spectrum of variation would become rather large.

If people could have traveled easily, these variations would have been distributed more evenly throughout the world. But since populations were isolated by large distances, each group's unique variations remained preserved. If one adventurous group living in northeast Africa then chose to venture forth into the Middle East, say 100,000 years ago, their descendants would be able to trace their origins back to these first out-of-Africa migrants based on the shared variations in their genomes.

It is important to note, however, that the groups that left the continent carried within them only a portion of the full spectrum of African genetic diversity that existed at the time, and it was these subsets that were passed on to subsequent generations in other parts of the world.

Understanding this genetic "sampling" is as easy as grabbing a handful of jelly beans from a large bag of mixed flavors. If the bag were large enough and the variation extensive enough (imagine thousands of flavors), it would be very difficult to capture the true complexity of flavors available by examining only a small sample of jelly beans.

Similarly, human migration out of Africa essentially sampled the greater African genomic diversity, and this pattern repeated itself as humans migrated across the globe. The genetic diversity that arose in humans in Europe and Asia varied geographically. When humans crossed the Bering Strait into the Americas, for example, they carried genetic patterns more similar to those of East Asians than of Europeans (see Figure 8).

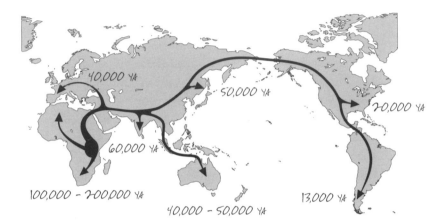

FIGURE 8: MAP OF EARLY HUMAN MIGRATION
Although the exact chronology of early human migration is still being determined,
the general consensus is that modern humans emerged in Africa between 100,000
and 200,000 years ago and began venturing out of the continent some 60,000 years
ago, with groups eventually spreading across the globe.

Y-Chromosomal Adam

Since the human species had to start somewhere, it is easy to imagine a "first" man and woman from whom we are all descended. Genomics provides us interesting ways of looking back in time to see when and where these individuals lived. The answers we find, however, might be surprising.

The key to looking back along our male lineage is the unique DNA that only about half of us carry—the Y chromosome. Anyone who inherits a Y chromosome is a man, and if a man fails to have a son, he will not pass on his Y chromosome. This means that a particular Y chromosome can become "extinct" from future generations. This happens largely by chance and usually only when population sizes are rather small.

A useful way to illustrate the process is by examining another "trait" generally passed on by members of one gender: surnames. An interesting story of last names becoming extinct in a small population involves the Pitcairn Islands, a group of four volcanic islands in the southern Pacific Ocean that were uninhabited when six mutineers from the famed British ship *HMS Bounty* and the 13 Tahitian women traveling with them settled the main island of Pitcairn in 1790. More than 200 years and approximately seven generations later, the population of the islands is only about 50 people, with only four surnames shared among them—three from original *Bounty* crewmen and one of a whaler who later settled there. The loss of half the original names is due to chance "extinction" events that occur when men do not have sons. If the population of the Pitcairn Islands remains small, it is likely that only two, or even one, of the original surnames will survive in the future.

What happened with male surnames on the Pitcairn Islands is akin to what happened with countless Y chromosomes over the course of human history. A number of researchers who analyzed Y chromosome DNA of males from all regions of the world have similarly concluded that the Y chromosomes of all males can be traced back to a single male who lived in Africa, probably between 60,000 and 90,000 years ago. This *Y chromosome most recent common ancestor (Y-MRCA)* is often referred to as "Y-Chromosomal Adam," a reference to the biblical Adam.

Whereas the name might suggest that this person was the "first human man" and the only living male of his time, this is far from the truth. Y-Chromosomal Adam was one of many living human males. It is just that the other Y chromosomes were lost over time, leaving us with no direct, unbroken male line traceable to a contemporary of Y-Chromosomal Adam. Either the lines of the other males died out or at least one generation within each line produced only daughters who, obviously, had no Y chromosomes to pass on.

Scientists have made these inferences using a parsimony-based analysis. Not all males alive today have identical Y chromosomes; but the mutations that occurred over time can be traced across generations, enabling us to reconstruct both when and where our oldest common male ancestor lived.

MITOCHONDRIAL EVE

Although the X chromosome might seem a logical tool for tracing maternal lineages, the fact that a woman passes on her X chromosomes to her offspring and that a man can also pass an X chromosome to his daughter makes using the X chromosome for tracing female lineage impossible. To do so, we must look outside our definition of the human genome as being the collection of DNA in the cell's nucleus, and instead look to a unique piece of DNA passed on only through the egg—*mitochondrial DNA (mtDNA)*.

The mitochondrion is a remarkable creation in the evolution of life on the planet. This organelle functions as a biological "power plant," producing adenosine triphosphate (ATP), the cell's main source of chemical energy. Mitochondria arose very early in the evolution of eukaryotic organisms. It is believed they were once independent, parasitic *proteobacteria* that established an endosymbiotic relationship with their host cell by taking up residence within the cytoplasm and incorporating themselves into the cell's metabolic processes. An internal source of energy is highly advantageous for a cell, and this explains why mitochondria exist in nearly all known eukaryotes.

Not surprising given its origins, the mitochondrion has its own independent circular genome, which in humans is 16,569 base pairs long (see Figure 9). More relevant to this discussion is that mitochondria are passed from a woman to her offspring through her egg cells, and since male sperm cells do not contain mitochondria, each of us carries only our mother's mitochondria.

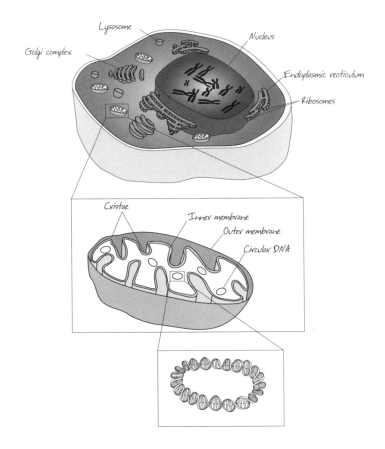

FIGURE 9: MITOCHONDRIA AND MITOCHONDRIAL DNA

Mitochondria are organelles with a smooth outer membrane. Their inner membrane is highly convoluted with folds called cristae. Cristae increase the surface area of the inner membrane, where sugar is combined with oxygen to produce ATP, the primary energy source for the cell. Like the bacteria from which they evolved, mitochondria can carry many copies of their circular genome.

If a woman has no daughter, her mitochondria will not be passed on to future generations.

What makes mitochondrial DNA unique and particularly useful is that it does not undergo recombination like the chromosomal DNA in the nuclear genome (remember the LEGO constructions?). Your mitochondrial DNA is essentially identical to your mother's and to your maternal grandmother's, and so on. Only mutation can introduce changes into mitochondrial DNA, although it rarely happens. By looking at many people across generations, we can estimate this mutational rate. This means that mitochondrial DNA can serve as a "molecular clock," whereby the number of mutational changes can serve as an approximate measure of generations—and hence an approximate measure of time.

By sequencing mitochondria from people around the world and performing data analysis similar to that done on the Y chromosome, scientists have traced back all living humans to a woman who lived in East Africa between 150,000 and 250,000 years ago. This *mitochondrial most recent common ancestor (mt-MRCA)* is referred to as "Mitochondrial Eve."

Passed down through the ages from mother to child, derivatives of "Eve's" mitochondrial DNA are found in all living humans. As with Y-Chromosomal Adam, Mitochondrial Eve was not the only member of her sex alive during her time. Rather, the mtDNA of her contemporaries simply did not survive over the millennia. Of course, the women alive at the time of our Mitochondrial Eve had their own mitochondrial most recent common ancestor—their own Eve.

To understand this, it helps to clarify that Mitochondrial Eve and Y-Chromosomal Adam represent our most recent female and male common ancestors. If a disaster occurred tomorrow and only a small isolated population of our world survived, it would would represent only part of the Y chromosome and mitochondrial genetic variation that existed on earth the day before. The most recent common male and female ancestors of these survivors would be much more contemporary than the ancestors we have identified for today's entire human race.

For example, if you and your siblings became the last humans on earth, your parents would be considered the most recent common ancestors, even though human beings have been around for much longer. If you found out that cousins born to your mother's brother were also alive, then your maternal grandparents would be most recent common ancestors. As you can see, determining who is the most recent common ancestor involves understanding the relationships between who is alive now and whose DNA survived thousands of generations of human life on earth.

Some might wonder why a huge separation in the time exists between when our male and female most recent common ancestors lived (60,000 to 90,000 years ago for Y-Chromosomal Adam versus 150,000 to 250,000 years ago for Mitochondrial Eve). The answer is partly based in culture. Throughout human history, fertile women have had more or less equal chances among one another of giving birth to a number of fertile descendants; but the chances among fertile men tended to vary more significantly, with some fathering no children and others fathering many and with multiple women.

NEANDERTHALS: OUR LOST COUSINS

Nearly everyone has heard of the Neanderthals (*Homo neanderthalensis*), a species closely related to *Homo sapiens*. Fossil evidence indicates that the Neanderthals lived in Europe and parts of western and central Asia beginning about 130,000 years ago, then rather suddenly disappeared about 30,000 years ago. Many have wondered if Neanderthals were humans to begin with and whether they simply went extinct or instead interbred with expanding tribes of *Homo sapiens* to become absorbed into the modern human species.

In February 2009, scientists from the Max Planck Institute announced their completion of a first-draft sequence of the Neanderthal genome using DNA extracted from the thigh bone of a 38,000-year-old male found at Vindija Cave in Croatia and sequenced along with the DNA of other bones from Spain, Russia, and Germany. This draft genome sequence, which was added to genomic sequence data generated by other groups, helped answer questions regarding our relationship to this very humanlike species.

At roughly 3.2 billion base pairs, the Neanderthal genome is about the size of the modern human genome. Data indicate that human and Neanderthal DNA are 99.5 percent identical (compared with about 96 percent similarity between humans and chimpanzees). However, no DNA evidence suggests that humans and Neanderthals interbred; instead it seems the two share a most recent common ancestor who lived between 300,000 and 500,000 or more years ago. This implies that in terms of evolutionary time, Neanderthals were not a branch of modern humans who left Africa before we did, but a distinct cousin species that went extinct about the time that humans expanded into their territory.[1]

ARE WE THE THIRD CHIMPANZEE?

While most of us have heard of Charles Darwin's *The Origin of Species*, not everyone is familiar with his second great work, *The Descent of Man*, published in 1871. In it, Darwin was one of the first to suggest that humans originated in Africa and that we are related to the great apes who inhabit that continent: "In each great region of the world the living mammals are closely related to the extinct species of the same region. It is therefore probable that Africa was formerly inhabited by extinct apes closely allied to the gorilla and chimpanzee; and as these two species are now man's nearest allies, it is somewhat more probable that our early progenitors lived on the African continent than elsewhere."

Darwin was extraordinarily prescient in proposing this hypothesis, but it remained largely speculative until the 1980s and 1990s, when it was confirmed by evidence accumulated through genomic analysis and the fossil record evidence of ancient specimens. However, in addition to clearly indicating that humans evolved in Africa then spread worldwide, genomic evidence has also revealed a profound anthropological surprise involving our place in the evolutionary tree.

By simply comparing physical characteristics, it was logical for Darwin and others to speculate that humans share an ancestry with other apes. Our membership in the biological order of primates is based on the characteristics we share with them.[2]

However, the general consensus among scientists had been that humans were a somewhat special type of primate, and that the great apes, including chimpanzees, gorillas, and orangutans, diverged from us at least 15 to 30 million years ago. This, however, does not appear to be the case.

Humans and the two chimpanzee species—the common chimpanzee and the bonobo—are in fact more closely related to each other than to the other great apes. Furthermore, humans split off as a separate species only about 4.7 million years ago (see Figure 10).[3]

A series of articles published in the journal *Science* in 2009 described the extensive analysis of a 4.4-million-year-old fossil called "Ardi" (for *Ardipithecus ramidus*). Discovered in 1992, its significance has only recently been recognized. Extensive analysis of Ardi showed that although chimpanzees and humans share a common ancestor, chimps are not primitive humans. Instead, chimp species evolved along different paths. For example, chimp feet are specialized for grasping tree branches, but Ardi's feet are better suited for walking.

In 2005, a draft genome sequence of the chimpanzee was published by scientists at the NHGRI, and its comparison with the human genome sequence led to some interesting findings. The analysis identified about 600 genes, the sequences of which have diverged significantly since humans and chimps took separate evolutionary paths from their common ancestor. These include many genes involved in immune system defense against microbial disease and pathogenic microorganisms. However, nearly half of the 600 are *transcription factors*, master regulatory

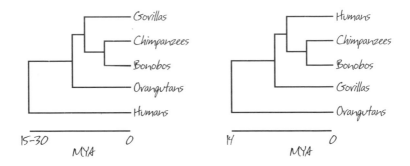

FIGURE 10: PHYLOGENY OF THE GREAT APES

Prior to the availability of genomic technologies, the great apes (orangutans, gorillas, and both bonobos and common chimpanzees) were thought to comprise a phylogenetic group separate from that of humans, who had diverged from the apes approximately 15 to 30 million years ago (left). However, mitochondrial DNA studies and DNA sequencing indicate that bonobos and common chimpanzees are actually more closely related to humans—and humans to these chimpanzee species—than they are to gorillas (right). Humans are also believed to have diverged from the chimpanzees much later (approximately 4.7 million years ago).

genes that control when and where other genes are turned on and off. Changes in these genes might account for the greatly differing phenotypes of humans and chimpanzees, despite the high degree of similarity in their genotypes.

Among the genes that differ between humans and chimpanzees are several genes involved in hearing, as well as FOXP2, a transcription factor implicated in speech development, suggesting that one of the critical factors separating us from our closest relatives is the development of language.

Also worth noting is that the human and chimpanzee ancestors that lived 4.7 million years ago were different from the humans and chimpanzees that exist today. Those protohumans were very apelike in their appearance. Our lineage underwent innumerable changes and evolution conducted many experiments, including ones that led to genetic dead-ends, before anatomically modern humans emerged 200,000 years ago.

Genome sequencing continues to yield valuable data about ever more species, as well as surprises about our evolutionary connection to those species closest to us in the tree of life. As for who we are as humans, the technology has already given us the gift of knowing that despite all the conventions people have used to define the differences among them—race, ethnicity, gender, and others—all people are more similar to one another than anyone has imagined. At our genomic core, we are all Africans—or, at least, we were some 70,000 years ago.

CHAPTER 6 NOTES

1. A species is commonly defined as a group of organisms capable of interbreeding, but which is separated from other such groups by the normal lack of interbreeding. By this definition, the lack of genetic evidence of interbreeding between Neanderthals and humans supports the hypothesis that Neanderthals were a distinct species.

2. Characteristics that define primates include, among others, forward-facing eyes for good binocular vision; a shortened snout that contains at least three types of teeth (incisors, canines, and molars), reflecting a diverse diet; collarbones; fingernails and toenails instead of claws; two separate bones in the forearm (radius and ulna) and leg (tibia and fibula); and an enlarged brain with larger dedicated regions for seeing and smaller dedicated regions for smelling.

3 For perspective, life has existed on earth for 3.8 billion years, mammals have been around about 510 million years, and primates are thought to have appeared 65 million years ago.

CHAPTER 7

⸙

A Brave New Genomic World

From what has been discussed so far, it might appear that genomics has only made an impact on university laboratories, pharmaceutical and biotechnology companies, and people suffering from a handful of diseases for which there are genomic screening technologies. But the truth is that genomics has profoundly altered many areas of our lives in ways that have direct implications for our welfare.

Looking beyond our own species, genomics has increased our insight into much of the living world. More than 5,000 organisms have had, or are having, their genomes sequenced, and many more will undergo at least partial genome sequencing. And nearly all of them are important to us in one way or another. They include agriculturally important crops like rice, maize, wheat, soybean, and wine grapes; cattle, sheep, and other livestock; geographically and evolutionarily interesting species such as the kangaroo and duck-billed platypus; and disease-causing bacteria and parasites along with the insects and other species that transmit them. Sequencing projects are also targeting bacteria species that can remove heavy-metal contamination from water sources, archaeal species that live in the most extreme environments on earth, and ancient yeast strains that have been extracted from amber and revived (and used to brew beer). The list goes on, encompassing a wealth of diverse avenues of exploration.

In this closing chapter, we will briefly examine a few areas of human health and behavior that are being shaped by genomic advancements and what they might mean for us in our day-to-day lives and for our future. Although what follows is not a comprehensive survey or an in-depth analysis of topics, each of which could easily fill up its own chapter, what is discussed will illustrate genomic technology use at its most creative—and controversial.

FORENSIC GENOMICS

It was not long ago when blood was the best evidence a detective could discover at the scene of a crime. Blood type (A, B, AB, or O, with accompanying positive or negative Rh factor) was often the key piece, or only piece, of information available for determining guilt. The problem with relying on blood type, however, is that it is not a sure thing. Even the rarest type, AB-negative, accounts for about two million people in the United States. Clearly, blood type cannot uniquely identify an individual. In contrast, the human genome provides a much more specific personal identifier, one that can be applied in many ways, including in forensics and paternity testing.

Genomic testing actually dates back to before the Human Genome Project, when the variability between individual genomes was first explored. In 1985, at the University of Leicester in the United Kingdom, Alec Jeffreys discovered the existence of short, repeating DNA base sequences that differed among individuals. These microsatellites, or *simple sequence repeats (SSRs)*, are polymorphic series of DNA sequences one to six bases long that repeat multiple times at specific locations within the genome.[1] Jeffreys found that, taken together, multiple SSRs could serve as a molecular "fingerprint" that varies among people much more dramatically than characteristics like blood type.

Scientists soon discovered that sets of SSRs could be used for forensic genomic identification. The U.S. Federal Bureau of Investigation maintains the Combined DNA Index System (CODIS), the world's largest DNA database, housing well over five million DNA profiles created by local, state, and federal crime laboratories. The use of CODIS and similar databases maintained by other countries to identify crime suspects has become common practice.

Despite their value, DNA fingerprints are not unique because they survey only a few of the three billion bases in the genome. Additionally, their value in identifying a person in a criminal investigation has sometimes been overstated. It is easy to exclude someone from a crime based on DNA evidence; inferring guilt is another matter entirely.

This can be illustrated by looking at another use of DNA-based SSR markers: paternity. If a woman sought to determine which of two men fathered her child, then using forensic SSR markers to compare the child's DNA profile with those of the mother and the two men would provide persuasive evidence for identifying the biological father. This is because other strong evidence likely exists to suspect both men. The question here would not be, "Of all the people in the country, who is the father?" but rather, "Of the two men we know could likely be the father, whose DNA is consistent with the DNA evidence?" To answer the latter question, additional evidence—such as the woman's sexual history with the men—would help implicate the correct man. The DNA is being used to confirm a suspicion.

The same general principle applies in criminal forensics. DNA fingerprints can be highly discriminating, but not absolute. A set of markers might be carried by only one person in five million, but that still means there are at least 60 people in the United States sharing the same DNA fingerprint. In other words, absent other evidence, the odds of identifying the right person are only one in 60.

What poses a greater concern is that since DNA fingerprint patterns are not random—their frequency is correlated with race and ethnicity—it could be more likely that the specific pattern being traced would appear in a person who matched a witness description. Imagine, then, that you slightly resembled this physical description and carried the same DNA fingerprint as one collected at a Chicago crime scene, but you were in London during the commission of the crime. You certainly would not want that DNA fingerprint to be the factor driving suspect identification and trial. Therefore, if DNA evidence is to be used effectively, it should be one of many factors that link a suspect to a crime—or a biological father to his baby.

Of course, it is only a matter of time before the cost and accuracy of DNA sequencing technologies improve and they become frontline tools in forensics. The entire genome would be a vastly superior unique identifier (allowing for identical twins, of course) and radically change the way DNA evidence is used. If an evidentiary genome sequence matched a suspect's genome, there would be little doubt as to accuracy. Again, however, the data would have to be considered in the context of other evidence to prove that the person whose DNA was at the scene was indeed the person who committed the crime.

GENOMES "R" US

Once upon a time, computers were esoteric instruments, relegated to academic, government, and industrial laboratories. Then came the birth of the personal computer, which revolutionized society, not only by changing the way we interact with one another, but by driving down the cost of computing across the board, pushing microchips into everything from household appliances and automobiles to surgical equipment and space shuttles. A similar transformation is taking place with public access to genomic technologies and the emergence of companies offering genome screening to consumers. This growing availability is forcing us to rethink what genomic information truly represents.

Personal genomics companies typically ask clients to mail in a sample of their DNA, usually collected from the inside of the cheek using a cotton swab. The DNA is extracted from the swab, then profiled using the same DNA microarray technologies used in research laboratory studies. Available direct-to-consumer (DTC) genetic tests range from assays for breast cancer susceptibility genes to genes for diseases like cystic fibrosis to genetic screenings that offer information about one's genetic ancestry. The cost for the tests based on DNA chip technology is generally only a few hundred dollars.

If you are interested in something grander, however, there are companies that will sequence your entire genome for a price, although if you wait long enough, that price could fit your wallet. Just a few years ago, the cost for sequencing a human genome was in the hundreds of thousands of dollars. In 2010, at least one company offered whole-genome sequencing as a DTC test for less than $20,000, while another company promised to do the same for $5,000. And with the astonishingly rapid advances occurring in technology, it will not be long before whole-genome sequencing falls below $1,000. At that price it is easy to expect the market for genome sequencing to grow to include those who want to go beyond exploring ancestry and possible disease risk to look for possible explanations or treatments of a rare disease, should they have one.

Despite growing excitement, a great deal of controversy surrounds DTC genomics. Caveats accompany the information that genome sequencing can provide. To begin with, consumers can be misled by invalid or erroneous results. And even if results are accurate, our knowledge of many diseases and their causes, particularly for complex diseases, is far from complete. Whereas one could argue that giving people the means to take ownership of and control over their personal genetic information is valuable, DTC genetic test results are largely provided without the involvement of a health-care professional, leaving data interpretation largely up to the consumer. Without professional guidance, some people might lull themselves into a false sense of security or experience undue alarm about their risk of developing a genetically linked disease. These and other reasons illustrate the need for adopting a cautious approach when making genomic information accessible to the broader public.

LEARNING FROM WATSON'S GENOME

In May 2007, James Watson was honored by the scientific community for being the first individual to have his genome sequenced.[2,3] A team from Baylor College of Medicine in Texas and 454 Life Sciences of Roche Diagnostics employed "next-generation" DNA sequencing technologies to complete a more than seven-fold coverage of all six billion bases of Watson's diploid genome in just about two months.[4]

In comparing Watson's genome sequence with the reference human genome, the team identified 3.3 million single nucleotide polymorphisms in his sequence. The vast majority appeared outside of genes; however, 10,654 SNPs were contained within genes, signifying changes that could cause small modifications in Watson's proteins relative to the "reference" set of human proteins.

The potential functions of the vast majority of these modified proteins were unknown, but 12 of the SNPs represented alterations that, if homozygous, could give rise to diseases or other
recognizable physical phenotypes. Ten of these dozen are considered highly penetrant, Mendelian-recessive, disease-causing
alleles; meaning that if you carry two copies of the mutated gene,
you have a very high chance of developing the associated disease. Four of the 10 were found to be homozygous, indicating
that Watson should, by all accounts, manifest the disease.

Within this last group were the genes for Cockayne syndrome, a rare congenital disorder characterized by growth failure,
impaired development of the nervous system, abnormal sensitiv-
ity to sunlight, and premature aging; and for Usher syndrome, a
rare disease typified by deafness and a gradual loss of vision.

But Watson is not deaf and blind and does not suffer from
Cockayne syndrome. This tells us that even if armed with a great
deal of information on a person's health, we still cannot rely on
the genome alone to accurately interpret genetic risk. Factors
such as lifestyle choices, diet, habits (such as smoking), and
exposure to environmental chemicals may be as or more important than genetic factors, and serve as a cautionary note about
the potential risks of misinterpreting genomic sequence data.

STEM CELLS

The human genome contains the blueprint for the building blocks
that make up each and every cell in an adult, but it all begins
within a single fertilized egg. An especially fascinating aspect

about the earliest stages of human development is how the "program" encoded in the genome plays itself out as the egg divides over and over and the resulting cells differentiate into the numerous specialized cell types that form the tissues and organs of the human body. Following birth, this process of cell differentiation continues on a smaller scale, as tissues renew themselves and wounds heal, with cells differentiating to form more specific cell types that make up the body's tissues.[5]

It was the scientific study of development, wound healing, and tissue regeneration that led to the discovery of stem cells. In humans and other mammals, two broad classes of stem cells have been identified: *embryonic stem cells (ES cells)* and *adult stem cells*. ES cells are *pluripotent*, meaning that when stimulated properly they can differentiate into almost any of the body's more than 250 cell types. Adult stem cells, of which there are multiple types, are partially differentiated, therefore *multipotent*. Although they can differentiate into a variety of cell types, they do not have the same potential to do so as ES cells. For example, hematopoietic stem cells can differentiate into the cell types found in blood, and neuronal stem cells can differentiate into nerve cell types; but hematopoietic stem cells cannot form nerve cells and neuronal stem cells cannot make blood.

Many people consider stem cells as a promising source for disease therapies, but they might not know that stem cells have also been implicated in a wide range of diseases, including cancer and diabetes. For both reasons, scientists are working to understand the properties of stem cells and the stimuli that cause them to differentiate. The greatest interest lies in the use of ES cells as potential therapies; however, until recently, progress has been hampered by ethical concerns, particularly in the United States.

In 1998, James Thomson and colleagues at the University of Wisconsin created the first human embryonic stem cell line that could be propagated in the lab and used to study ES cell properties. Building on Thomson's work, a small number of other human ES cell lines were quickly created, but concern arose over the use of in vitro fertilized embryos as the source of the ES cells. This resulted in a ban of federally funded research that used any but a small set of previously established ES cell lines. Unfortunately, many of the existing lines were subsequently found to be of poor quality.

Despite these limitations, genomic and other analyses of existing stem cell lines (including work on human ES cells conducted outside the United States), along with analysis of ES cells in mouse, rat, and other species, has revealed much about the genetic programming of stem cells. In Japan in 2006, biologists Kazutoshi Takahashi and Shinya Yamanaka showed that the introduction of just four expressed genes into fibroblasts—highly differentiated cells from which connective tissue develops—was enough to convert these cells into *induced pluripotent stem cells (iPSCs)* that possessed many of the apparent properties of ES cells.

Although there is much more to learn about stem cells, their broad range of potential therapeutic applications has generated a tremendous excitement in the biomedical research community. Hope rests on the possibility that a person's own cells could one day be used to create iPSCs for therapeutic use, and thus avoid the problem of rejection by the immune system and the ethical issues surrounding the use of ES cells. Regardless, many scientists remain cautious about iPSCs. Creating them requires activating pathways that are known to be activated in cancer and these pathways may remain active, putting cells at increased risk for becoming cancerous. This poses a significant problem if iPSCs are to be used to treat human diseases.

Moreover, the concern about the potential cancer risk of iPSCs has been amplified by the growing evidence that cancer development and progression—including tumor metastasis, therapy resistance, and disease relapse—are driven by *cancer stem cells (CSCs)*, believed to comprise a small subpopulation of tumor cells. Like other tissue-specific stem cells, CSCs can give rise to all the cell types found in their particular tissue that are necessary for a cancer tumor to develop, grow, and survive. For this reason, CSCs are considered to be *tumorigenic*, operating through the stem cell process of self-renewal and differentiation into multiple cell types.

Genomic tools offer the best way to investigate the molecular, biochemical, and functional nature of stem cells. In the case of CSCs, there is great interest in developing new therapies that target these cells based on their unique properties, such as the specific stem cell markers that reside on their surfaces. And in general terms, understanding how and why cells differentiate is critical not only to the study of cancer but to many other diseases, including diabetes, for which stem cells might be used to create the insulin-producing cells lost to the disease; or neurological disorders like Parkinson's disease, where stem cells could potentially create the body's missing dopamine-producing cells.

A Word About Cloning

For the more imaginative reader, the science of genomics might suggest the possibility of our one day being able to clone a wholly new organism that carries a person's exact genetic code. Doubtless, the prospect of "renewing" our bodies by growing replicates of ourselves or of somehow transferring our consciousness to clones sounds exciting. However, science fantasy must give way to scientific fact, and while this avenue of inquiry does not have much to do with genomics, per se, it is worth at least a brief note here.

Most everyone has heard of the famous cloned sheep named Dolly, the first mammal (although not the first animal) to be cloned. This feat was accomplished in 1996 by Ian Wilmut, Keith Campbell, and others at the Roslin Institute in Scotland through *somatic cell nuclear transfer (SCNT)*. SCNT involves transferring the nucleus of an adult cell into an unfertilized *oocyte* (a developing egg cell) that has had its nucleus removed. Using electric shock, the hybrid cell is stimulated to divide, then transferred into the womb of a surrogate female. In Dolly's case, the cell nucleus came from a mammary cell (hence the sheep's name as an allusion to country music singer Dolly Parton).

Other mammals have been cloned using modifications of SCNT. In January 2008, Samuel Wood and Andrew French at the California company Stemagen Corporation announced the creation of five cloned human embryos using skin cells taken from Wood and a colleague. (The embryos were destroyed five days later.) The researchers' stated goal was to develop a process for obtaining viable embryonic stem cells for therapeutic cloning that could lead to disease treatments.

Regardless of its potential, therapeutic cloning raises a host of unresolved ethical issues, but they pale in comparison to the provocative questions attached to reproductive cloning, which produced Dolly. Would a person with a DNA sequence identical to yours be considered a separate individual and have his or her own right to life? Could you create a clone as a source of replacement parts? Could your clone even replace you entirely to give you a second shot at life?

Well, such clones are alive today in the form of the identical (*monozygotic*) twins nearly all of us know. Clearly, someone who shares your DNA is not you. Genetics might influence everything from temperament to food preferences, but it is life experience that makes people unique individuals. So even though there are many questions regarding cloning that cannot be simply answered, understanding them is important since the legal system is still wrestling with their implications.

As for replacement cloning, in which you would grow a "new you" and somehow transfer your consciousness to your clone, that scenario is best left to science-fiction writers.

PRENATAL GENETIC TESTING AND GENE THERAPY

During pregnancy, every prospective parent asks the same questions: Will my baby be born healthy? Will she or he grow up to become an athlete? A musician? A mathematical whiz? As our knowledge about the link between genotype and phenotype increases and as DTC genomic assays become more common-place, more opportunities will open up for genomics to affect fundamental aspects of our everyday lives, including family planning.

When it comes to determining the health of an unborn fetus, genetic tests have been available for many years. The most common and easiest to interpret prenatal test is the ultrasound, used routinely to determine gender, but also to detect structural and anatomical abnormalities. Other tests can spot large-scale chromosomal abnormalities, such as trisomy 21, trisomy 18, and trisomy 13 (introduced in chapter 1). These tests involve obtaining cells from the developing embryo—such as through amniocentesis, whereby fetal cells are extracted from the amniotic fluid in the womb—and microscopically analyzing them to inventory the chromosomes.[6]

As the spectrum of known mutations and their links to disease grows, so will the availability of modern molecular biology tools to examine specific genomic alterations. Currently, the autosomal recessive disorders most frequently diagnosed using prenatal genetic screens are cystic fibrosis, beta thalassemia, sickle cell disease, and spinal muscular atrophy (type 1). The most common autosomal dominant diseases diagnosed are myotonic dystrophy, Huntington's disease, and Charcot-Marie-Tooth disease. For X-linked diseases, where mutations are found on the X chromosome, most of the tests performed are for fragile X syndrome (a form of mental retardation), hemophilia A, and Duchenne muscular dystrophy. It cannot be long before we can scan the entire genome for potentially deleterious mutations or other gene variants.

Medically, there are three primary reasons why prenatal screening can be important. First, it might enable timely medical or surgical treatment before or after birth. Second, identifying a particular genetic abnormality allows parents to make informed decisions about whether to go forward with a pregnancy. And third, it gives parents the chance to prepare mentally, financially, and medically for a baby born with special needs or for the likelihood of a stillbirth.

Prenatal genetic testing does raise serious moral and ethical issues for some people who must decide whether or not to terminate a pregnancy. Would it be justified if a genetic disorder such as Down syndrome were detected? What if it were discovered that a baby would be born with severe congenital myasthenic syndrome, a rare genetic condition in which the sufferer cannot breathe on his or her own? There are no easy answers here, either for parents or for policy makers, and exploring these considerations lies outside the scope of this book.

There is, however, another potential use for testing that raises particular concern: the ability to plan for what some might call "designer children." In 1993, scientists at the Institute for Neuromuscular Research at the Children's Hospital at Westmead in Sydney, Australia, found that variants of a gene called ACTN3 were more prevalent in elite athletes than in the general population, suggesting that the gene plays a role in athletic performance. An Australian company called Genetics Technologies began offering a simple genetic test for the presence or absence of this "athletic prowess" gene. And although ACTN3 is not the sole determinant of one's athletic ability, a great deal of controversy surrounded the test. Would parents use it to test their kids and determine which of them should take up sports? Would athletic recruiters at high schools and colleges begin testing potential athletes?

This is just one example of the kinds of questions we might ask in the future as we discover more and more genes associated with physical and even psychological characteristics that are not disease-related. How will we manage and use this information? Gender is already a factor considered in some situations when determining whether to continue a pregnancy. If they could, would parents request prenatal tests during in vitro fertilization to select a "preferred child" from a collection of fertilized embryos? Taking this logic one step further, will we one day be able to genetically engineer not only what we are but who we are? It is important that everyone considers how this genetic information and testing could, and should, be used in the future.

One question that comes up with genetic testing, particularly prenatal genetic testing, is whether something can be done if a mutation is discovered. Correcting a mutation by introducing a corrected version of the gene into each and every cell seems, at face value, to be an attractive idea. However, there are problems associated with gene therapy, including the tremendous difficulty in transferring genes efficiently into cells and the potential disruptions that transferred genes might cause for other genes. For these and other reasons, the science of gene therapy is still in its infancy despite years of scientific work. Although limited successes have motivated additional research, gene therapy is unlikely to serve as a viable approach for the vast majority of human diseases for the foreseeable future.

LEGAL AND ETHICAL CONCERNS

Given the wealth of genomic data being generated ever more rapidly and easily, we all might one day have to decide whether we want our genomes sequenced. While the prospect certainly sounds exciting and promising, particularly with regard to disease prevention and treatment, the potential risks in having one's DNA sequence made available should not be ignored. It is important to keep sight of the ways in which we, and society in general, will treat such profoundly personal information.

Possible misuses are many. Anyone with sufficient knowledge could use your genome sequence to infer paternity or other features of your genealogy; claim statistical evidence about your risk for diseases or other characteristics that could affect obtaining employment, insurance, or financial services; reveal the propensity for your developing a disease that currently lacks effective treatments; claim your relatedness to criminals or incriminate your relatives; or even make synthetic DNA that corresponds to your DNA and plant it at a crime scene. Granted, it is easy to dismiss a couple of these scenarios. For example, it would be far simpler to steal your comb and leave strands of hair at a crime scene than to forge your DNA sequence; and inferring paternity and genealogy does not require access to your entire genome sequence.

Nevertheless, information on your genetic propensity for developing disease could be misused. Insurance companies operate on the principle of shared risk and often exclude patients from coverage if they have a preexisting condition that is expensive to treat. The question naturally arises of whether certain gene variants would constitute a basis for denying coverage. We know, after all, that mutations in the APOE gene are associated with increased risk for Alzheimer's, a risk not equally shared across the population. If you carried mutations in the APOE gene, should your insurance company be allowed to exclude you from coverage for Alzheimer's disease? Similarly, if employers had access to genomic information, could they use it to decide whom to promote or send off for expensive training? After all, why should they invest in someone at increased risk for breast cancer or schizophrenia or heart disease?

Providing consumers with protection against the misuse of genetic information has been addressed to some extent by the U.S. government. On May 21, 2008, the Genetic Information Nondiscrimination Act (GINA) was signed into law, barring health insurance companies and employers from discriminating against individuals on the basis of their genetic information. This is a good initial step; but as genomic data becomes cheaper and easier to obtain, other questions and problems associated with its use will certainly arise and require addressing.

OUR GENOMIC FUTURE

The desire for invention and innovation has always been a driving force in human progress, particularly in science, where advancements are heralded as positive contributions for us and for the world in which we live. However, nearly every technology, no matter how beneficial, brings with it the potential for misuse. We have seen this unfold throughout the course of our history.

At the dawn of the twentieth century, a new understanding of physics, based on quantum mechanics, and the powerful tools that it yielded transformed our world in ways that no one at the time could have imagined, and in some ways it might not have been to our benefit. Quantum mechanics led to the development of the transistor, which eventually gave us the integrated circuit and the proliferation of computers in most homes and workplaces, as well as in items from toasters to car engines to greeting cards. But quantum mechanics also led to the birth of the atomic bomb and our current worries about nuclear proliferation.

The dawn of the twenty-first century marked an equally profound transition in our understanding of our genetic heritage. Within only several years, we went from ignorance about the human genome to completing its sequence of three billion bases. Genomic technologies are now reaching into every corner of our lives, and the implications might not always be fully evident, at least not yet.

Genomics provides us with a better understanding of human disease and potential new cures, but its application is also raising questions about genetic selection, or genetic engineering, of individuals with particular traits. Genomics promises to change agriculture and the raising of livestock, allowing us to produce more drought-resistant crops and cattle that provide more milk. Yet concerns surround how genetically modified organisms will ultimately affect the environment and the genetic diversity of the species we already rely on. Genomic technologies may lead to the development of synthetic life that is engineered to perform certain tasks or to new biologically driven methods for producing drugs or other important compounds.

With a bit of imagination, one could even see the impact of genomics in the social arena, with genomic profiles replacing questionnaires on Internet dating sites or DNA sequences posted on genomic equivalents of networking sites like Facebook or MySpace. As wide as the possibilities are for genomics to improve our lives (or at least make them more interesting), so are the risks in how our genomic data and technologies might be misused.

Now that we have unlocked the DNA from inside our cells and are reaping the extraordinary volumes of information it contains, it is hard to predict just where all of this ever-unfolding knowledge will take us and what its ultimate impact on humanity will be. But as in the myth of Pandora's box, the genome has been released and it has become a part of our world, and so we must be prepared for the changes it will bring. Whatever transpires in the next century, we must always remember that the genomic future is one that we each hold in our hands.

CHAPTER 7 NOTES

1. One of the most often used SSRs are CA-repeats, stretches of DNA sequence, in which a number of CAs follow each other (CACACACA, and so on). CA-repeats are easy to find in the genome and are highly variable among individuals. A technique called *polymerase chain reaction (PCR)* allows these repeats to be "amplified" from a DNA sample that includes the surrounding sequence. The number of repeated CAs is then determined by simply measuring the length of the amplified DNA fragment.

2. You can explore Watson's genome sequence online at the Cold Spring Harbor Laboratory (http://jimwatsonsequence.cshl.edu/). Interesting to note is the deliberate "hole" left in the genome sequence for the apolipoprotein E, or APOE, gene, the status of which Watson did not wish to know because its mutation predisposes a person to Alzheimer's disease.

3. J. Craig Venter sequenced his own genome about the same time, marking the first instance of a "vanity" genome sequence. His genome sequence was actually published before Watson's, raising the question of whose genome was really the first to be sequenced.

4. The diploid genome represents the entire sequence of both copies of Watson's chromosomes, which is twice the size of the haploid genome we refer to as "The Human Genome." The deep coverage is necessary to ensure that nearly every base is represented in the data and reflects statistical sampling theory. With one-fold coverage in randomly selected sequences, only about 63 percent of the genome would be represented; three-fold coverage would represent about 95 percent of the bases.

5. Although mammals have lost their ability to completely regenerate limbs, other species such as salamanders can. In general, the more complex an organism is, the less capable it is of regeneration. Humans do have the ability to regenerate one tissue—the liver, which can grow back from as little as 25 percent of its original tissue.

6. Cytogenetic testing is only used on chromosomes 13, 18, and 21 because most other large-scale chromosomal abnormalities are embryonically lethal; they disrupt development so severely that the fetus spontaneously aborts, often without the woman ever knowing that she was pregnant.

AFTERWORD

⋚

As a young boy, I was fascinated by science. Although I always sided with heroes like Batman and Underdog, I held a secret admiration for mad scientists who, with a secret formula, could change the world (though not always for the better). My early attempts at science involved crazy experiments, like trying to create my own secret formula by mixing cleaning products in the bathtub (something I now recognize as being dangerous). Despite my early misadventures, I remained certain that science was what I wanted to do. My journey from bathtub chemist to genome scientist, however, was atypical and reveals interesting parallels between my personal history and that of the genome project.

In 1990, I was completing my doctorate in theoretical physics at UCLA and looking forward to a very promising career. I had already done interesting work in cutting-edge areas of physics and had even won a few prizes. I had also secured an offer for a postdoctoral fellowship with a leading theoretical physics group. But forces beyond my control would throw a monkey wrench into my well-laid plans.

As described in chapter 1, the end of the Cold War brought about big changes in the funding for basic physics research. Physicists had long-before made the error of promoting physics as being vital for maintaining the United States' lead over the Soviet Union in matters of defense. (Personally speaking, no physicist I knew did anything even remotely related to defense.) When the Berlin Wall was torn down and the Soviet Union began to fall apart in the late 1980s, funding for physics, including my promised postdoctoral support, dried up. Although I took on temporary research positions for the next two years, I could read the writing on the wall.

Fortunately, I had a number of friends in the biology department at UCLA and had helped many of them solve thorny research problems by applying mathematical and problem-solving skills acquired through my physics training—skills that allowed me to look at their research questions in ways they did not. I soon learned that biology was undergoing a revolution as profound as that which had occurred in physics in the late 1800s and early 1900s, when novel experiments and new methods of investigation set the stage for people like Niels Bohr, Erwin Schrödinger, and Albert Einstein to dramatically alter our understanding of the universe. What I found exciting was that, in a similar way, new methods and experiments in biology were providing data for reimagining the way we understood ourselves.

When the U.S. Department of Energy (DOE) and, later, the National Institutes of Health (NIH) began the Human

Genome Project (HGP), they knew the project would require an approach unlike that of traditional biological research, in which small groups of scientists worked on single genes. Instead, they enlisted large teams and undertook an international effort associated more with large collaborative experimental physics studies. They also realized that they needed individuals with skills, including mathematical skills that differed from those of most biologists.

So the DOE and NIH decided to offer research fellowships to nonbiologists. For me, this was the classic no-brainer. Thanks to what I had learned about biology and the emerging science of genomics from my colleagues, I already possessed the basic knowledge required to understand the genome project, and the biology problems I helped solve gave me the credibility I needed to compete for a fellowship. As a result, I was in the first group of NIH fellowship applicants and was awarded five years of salary and research support to transition to genomics.

From the start, I realized that to be a leader in my newly chosen field, I had to go beyond my mathematical and computational skills and learn the fundamentals of molecular biology and genetics, as well as the new laboratory techniques and technologies that were driving the field forward. I began in 1992 at the Salk Institute in La Jolla, California, working on a project that involved mapping genes (and other landmarks) along the genome. Two years later I moved to the Stanford Human Genome Center (at Stanford University) in Palo Alto, where I led a group that sequenced regions of human chromosomes 4 and 21, associated, respectively, with Huntington's disease and Down syndrome. In

1997, I moved to The Institute for Genomic Research (TIGR) in Rockville, Maryland, to build a program focused on DNA microarrays and gene expression analysis. Since 2005, I have been at the Dana-Farber Cancer Institute and the Harvard School of Public Health in Boston, where I try to leverage all that we have learned from the HGP and the technologies it enabled so we can explore the basis for human disease and develop new medical tests and treatments.

When I sat down to write this book, I thought back to the days of my transition from physics to genomics. The fundamental questions I asked myself early on were, "What is a gene and why are genes so important?" I realized that anyone interested in learning about the human genome had to start at the same place. Of course, these questions led to more questions, invariably prompting me to consider everything from the history of cellular biology and genetics to how the human genome was sequenced to the ways that genomics helps us diagnose and treat disease and the role that genomic information plays in our everyday lives. In short, I decided I would tell the story as I have seen and lived it.

In a very real sense, however, the story of the human genome is a never-ending work in progress, and examples of this are everywhere and here to stay. On June 8, 2010, the genomics instrument company Illumina announced it had reduced the cost of its direct-to-consumer whole-genome sequencing service to $19,500, a price that will doubtless continue to fall. Yet even the CEO of Illumina has conceded that interpreting the data produced will likely be a challenge for years to come.

Maybe one day the cost will fall low enough and the data interpretation will be good enough that I, too, will have my genome sequenced—or maybe not. What *is* certain is that in our lifetimes, or our children's lifetimes, genomic science will play a major role in nearly everything we do.

And to say "nearly everything" is no exaggeration. Although I've made genomic science my life's work, I continue to be surprised by its increasing impact on the real world. Recently, while working in Brisbane, Australia, I took advantage of the city's impressive and well-run public transit system and hopped aboard a local bus. Brisbane prides itself on the cleanliness of its buses and is apparently using genomics to help maintain its standards. A sign posted on my bus read, "Spitting is unacceptable. Bus operators are now equipped with DNA kits to assist with the apprehension of offenders." Such an unexpected and creative application of DNA technology serves as a simple, yet powerful example of how the genomic future, with all its practical implications, is already upon us.

INDEX

adenine. *See* nucleic acids
AIDS (acquired immune deficiency
    syndrome), 50
    See also HIV
AlloMap, 135
Alwine, James, 96
Alzheimer's, 14, 174, 177
amino acids, 37, *38,* 40, 41
APC gene, 79, 94
APOE gene, 174, 177
Applied Biosystems, Inc. (ABI), 60
archaea, 28, 46, 52
Atomic Energy Commission, 16
autism, 58, 106, 128
Avery, Oswald, 33

bacteria
    E. coli, 48–49
    genome of, 68
    H. influenzae, 49–52
    M. genitalium, 52
bases. *See* nucleic acids
Baylor College of Medicine, 163
bioinformatics, 90
biology, molecular
    Central Dogma of Molecular
      Biology, 26, 136
Blair, Tony, 11, 13
Botstein, David, 101
BRCA1 gene, 84–85, 95
Brown, Patrick, 97, 101
Brown, Robert, 27

Cambridge University, 34
Campbell, Keith, 168
cancer
    breast, 14, 84–85, 101–3, 104,
      105, 106

causes, 91, 93–94
cervical, 68, 107, 108–10
definition of, 91
esophageal, 109
genes and, 93–95, 106–7
leukemia, 14, 99–101, *100*
lung, 130
lymphoma, 108
ovarian, 84
using genomics to cure,
    89–112
vaccines, 106
Cancer Genome Atlas, The
    (TCGA), 64
cardiovascular disease, 91, 106, 126,
    128, 129, 133
C. elegans (roundworm), 54–56
Celera Genomics, 14, 17, 18, 61–62,
    72, 73
cells
    early research of, 27–28
    eukaryotic, 28, 29, 39, 74
    prokaryotic, 28, 39, 46
    stem, 164–65
    types of, 86
cell theory, 28
Central Dogma of Molecular
    Biology, 36, 136
Charcot-Marie-Tooth disease, 170
chimpanzees, 67, 138, 151, 152–55,
    154
Christakis, Nicholas, 133–34
chromosomes
    bacterial artificial (BACs), 59, 61
    chromatin, 29
    crossover and recombination,
      116–18, *117*
    disorders in, 32, 170